Digestive Diseases in Sub-Saharan Africa

Digestive Diseases in Sub-Saharan Africa

Changes and Challenges

Isidor Segal
Previous Head, G.I.T. Unit, Chris Hani Baragwanath Hospital
and University of Witwatersrand, Soweto, South Africa

ELSEVIER
ACADEMIC PRESS
An imprint of Elsevier

Academic Press is an imprint of Elsevier
125 London Wall, London EC2Y 5AS, United Kingdom
525 B Street, Suite 1650, San Diego, CA 92101, United States
50 Hampshire Street, 5th Floor, Cambridge, MA 02139, United States
The Boulevard, Langford Lane, Kidlington, Oxford OX5 1GB, United Kingdom

Library of Congress Cataloging-in-Publication Data
A catalog record for this book is available from the Library of Congress

British Library Cataloguing-in-Publication Data
A catalogue record for this book is available from the British Library

ISBN 978-0-12-815677-3

Publisher: Mica Haley
Acquisition Editor: Stacy Masucci
Editorial Project Manager: Megan Ashdown
Production Project Manager: Poulouse Joseph
Cover Designer: Victoria Pearson

Typeset by SPi Global, India

Contents

Contributors

Monique Andersson
Department of Medicine, University of Cape Town, Cape Town, South Africa

J.W.S. Devar
Chris Hani Academic Hospital and University of the Witwatersrand, Johannesburg, South Africa

Suliman Salih Fedail
Chairman of Fedail Hospital, Consultant Physician and Gastroenterologist, Khartoum, Sudan

Alan Karstaedt
Division of Infectious Diseases, Department of Medicine, Chris Hani Baragwanath Hospital, University of the Witwatersrand, Johannesburg, South Africa

M.K. Kgomo
Head of Gastroenterology Unit, University of Pretoria, Pretoria, South Africa

T.E. Madiba
G I Cancer and Research Centre, University of Kwa-Zulu, Durban, Natal, South Africa

F. Makumbi
Specialist Physician-Gastroenterologist, Natal, South Africa

Hazel M. Mitchell
Professor of Medical Microbiology, School of Biotechnology and Biomolecular Science, University of New South Wales, Sydney, Australia

O. Mwantembe
Former Head of Gastroenterology, Steve Biko Academic Hospital, University of Pretoria, Pretoria, South Africa

S.R. Rutherford Thomson
St. Anne's Trust Chair and Head of Division of Medical Gastroenterology, Department of Medicine, University of Cape Town, Cape Town, South Africa; University College, London, United Kingdom

Isidor Segal
G.I.T. Unit, Chris Hani Baragwanath Hospital and University of Witwatersrand, Soweto, South Africa

Mashiko Setshedi
Department of Medicine, University of Cape Town, Cape Town, South Africa

M.D. Smith
Academic Head of Surgery, University of Witwatersrand, Johannesburg, South Africa

Mark W. Sonderup
Division Hepatology and Department of Medicine, University of Cape Town, Cape Town, South Africa

Frank Tovey
Department of Surgery, University College London, London, United Kingdom

Hester H. Vorster
Head of Centre of Excellence for Nutrition, Faculty of Health Sciences, North-West University, Potchefstroom, South Africa

C. Wendy Spearman
Division of Hepatology, Faculty of Health Sciences, Department of Medicine, University of Cape Town, Cape Town, South Africa

To my wife Arlene for her love and devotion
and to my dear children Rosh, Peretz, and Hadass
and their families for their continuous understanding
and support.

Foreword

The field of medicine has seen tremendous advances over the past few decades. This has been due, not in small part, to strong industrial- and institutional-funded research, largely in affluent nations of the western world. In the course of events, many prevalent diseases with significant morbidity and mortality in Afro-Asian countries became "orphan diseases." Afro-Asian physicians, dependent upon available western resources, were better equipped to handle western health problems than those of the home country. With improvements in worldwide institutions of higher learning, a rise in international cooperative studies, and better communication among immensely talented researches in the global community, medical advances have begun to penetrate Africa and Asia.

The challenges of third world diseases are still substantial. Only the clinician who has spent time in these settings can rightfully educate the student of the west. The flow of knowledge must be bidirectional. Just as the western world gained great insight into oral rehydration and intestinal absorption in the management of diarrheal disease from epidemic diseases in the third world, developing nations provide a fertile ground for good-quality research. Yet, due not only to socioeconomic biases but also to time limitations imposed by overwhelming clinical responsibilities, clinicians in developing nations continue to be challenged in their efforts to publish their work in high-impact journals.

Isidor Segal, MBBCh, PhD, FRACP, FRCP (London), is an eminent scholar, teacher, and clinical researcher. He has had decades of experience in Chris Hani Baragwanath Hospital Johannesburg, South Africa, the third largest hospital in the world, and offers to us the wisdom of many experienced clinicians of the continent. Decades of experience in this 3200-bed hospital and Witwatersrand Medical College is reflected in this book.

I have been fortunate to know Isi for nearly four decades, first in the field of pancreatology, and then in other clinical endeavors and publications. His innumerable publications on pancreatic diseases provided the best descriptions of the natural history of complicated acute and chronic pancreatitis. He published "Gastroenterology and Hepatology, a Clinical Handbook" (Churchill Livingstone) and a second book entitled "Gastroenterology and Hepatology Manual, a Clinician's Guide to a Global Phenomenon," which I was privileged to coedit with him. I have admired his wisdom in all areas of clinical medicine. Ii's humility often masks his enormous body of knowledge, but I am immensely pleased to see his readiness to share his knowledge with us all. Learning and teaching are lifelong efforts, and Isi has not wasted a minute in with either one. This monogram is a reflection of the wisdom of great teachers from the African continent.

The book covers various aspects of diseases of the emerging countries, along with the consequences of the acquisition of relative affluence. Westernization of diet and lifestyle, influx of alcoholism, results of industrial and environmental pollution, and other ecological changes are discussed. The book includes important topics such as peptic ulcer disease, H. pylori infection, parasitic diseases, inflammatory bowel disease, colorectal cancer, hepatocellular carcinoma, and others. With the enormous increase in African-Asian expatriates living in western nations, all physicians need to be familiar with geographic medicine, and benefit from the vast clinical experience of authors from emerging nations.

Isi, I am thoroughly impressed by your effort and I congratulate all of the authors on their outstanding effort. Your commitment and dedication to excellence is admirable.

I recommend this book to all medical students, residents, internists, and gastroenterologists who are interested in gaining a global view of diseases.

Professor C.S Pitchumoni

Chief of Gastroenterology, Hepatology and Clinical Nutrition, Saint Peter's University Hospital, New Brunswick, United States

Preface

"There is always something new coming out of Africa"
(Aristotle, 384–322 BC)

Globalization describes the way countries and the world interact always, "... *intimately associated with urbanisation*" [1].

Its economic web links the world as one large entity. Technology and communication systems have transformed people's lives.

However, a wind of despair is blowing across sub-Saharan Africa (SSA) where this transformation is being referred to as the "urbanisation of poverty" [1].

In SSA vast rural-urban migration is occurring, resulting in the creation of enormous slum areas. These poor living conditions embrace the establishment of shanty towns, inadequate housing, dietary changes, and poverty, concomitant with lack of clean water, primitive sanitation, and paucity of basic medical facilities [2]. The individuals caught up in this migratory vortex are further subject to the epidemic of HIV and the resurgence of tuberculosis. With regard to HIV, it is postulated that we are at last witnessing the prelude to gaining control of the scourge.

City life also means the clash of rural and urban cultures. Rural immigrants move into an environment, where they encounter the fragmentation of rural traditions and the breakdown of cultural norms. Furthermore, they are surrounded by powerful media messages without the technological experience to deal with them.

The inviting stimuli of convincing advertising found on radio and television have had a strong positive influence on increased smoking, eating junk food, and alcoholism.

City dwellers' patterns of drinking have also moved from traditional beer to commercial beers and spirits. Patients when questioned regarding the quantity of alcohol they drink often say: "*I drink on weekends and I drink until I drop.*"

The book identifies the relevance of the "poverty of urbanisation" [1] and the effects of this process on digestive diseases in SSA. It outlines the disease patterns that are emerging and investigates the possible solutions.

The focus is on diseases that are common in HICs and their manifestations in the "new world" of SSA. These chapters deal with a cross section of digestive diseases and are written by acknowledged experts in their fields.

The chapters present a continuous array of evolving conditions that include:

Pancreatitis, which underlines the devastating health hazard of alcohol; the increase in smoking of commercial cigarettes with dire consequences, and added to this mix, extensive environmental and nutritional changes that are being experienced in SSA cities. These are the harbingers of increase in obesity, diabetes mellitus, and cardiovascular diseases.

Interestingly, the paucity of some upper gastroduodenal diseases suggests that certain populations in SSA have protective factors that include the ramifications of *H. pylori*, believed to play an essential role in the "African Enigma."

While an exceptionally, elegant analysis on the influence of climate and duodenal ulcer has been identified in a unique study.

Amoebiasis and intestinal schistosomiasis soundly illustrate the diseases that are interwoven with poverty that will require strong political intervention to eradicate them.

It is intriguing that diseases that are common in HICs are also emerging in LICs as found through different observations. These various outcomes are discussed in chapters tracing the transition from traditional diseases to the present time.

However, there is a scarcity of medicines and modern medical equipment, and unfortunately SSA still lags far behind the HICs.

Despite this, there are regions in SSA where governance has improved and together with economic enlightenment there are already a few establishments regarded as centers of excellence. This gives hope that the vast subcontinent offers a ray of light to what will be the third most populous continent by 2050 [3].

The book is an important guide to doctors, students, interns, fellows, health care providers, and community groups and an invaluable reference for medical specialists in LICs. It is believed that the volition of the text will motivate action to bring about positive change in SSA.

REFERENCES

[1] Forbes D, Lindfield M. Urbanisation in Asia: lessons learned and innovative responses. Canberra: Australian Agency for International Development (AusAID); 1997. p. 1–886. Executive Summary, 10.247/BLT.09.073445.
[2] Patel DB. Urbanisation an emerging human disaster. N Eng J Med 2009;361(8):741–3.
[3] Ramin B. Slums, climate change and human health in sub-Saharan Africa. Bull WHO 2009;87:886. https://doi.org/10.247/BLT.09.073445.

Acknowledgments

My wife Arlene for her unstinting advice, patience, and fortitude in bearing the burden of an impatient husband.

Professors Steve Riordan, Joan Braganza, and the late Professors Leo Schamroth and Alexander Walker who taught me the essentials of research.

Grateful thanks to the following doctors who always encouraged my colleagues and me to develop the African Institute of Digestive Diseases: Meinhard Classen, Guido Tytgat, John Cummings, Chris Mulder, Eamonn Quigley, Godfrey Lule.

Stacy Masucci and Megan Ashdown for encouraging me and also to the Elsevier team for guiding me through the web of publishing this book.

I value the work all the contributors have done in support of the concept of the emergence of digestive diseases in Sub-Saharan Africa.

Introduction

The Gastro-Intestinal- Unit [GIT Unit] at Baragwanth Hospital was established in 1967.

One of the questions asked was: why a GIT unit in Soweto, since it was widely accepted that GIT diseases were uncommon in Black South Africans? Discussions with the pioneer of epidemiology in South Africa Prof. A.R.P. Walker confirmed this and suggested that it would be fascinating to investigate reasons for the rarity.

We were fortunate to have a comparison with the White South African population living adjacent to Soweto, who had a disease profile similar to that of populations in HICs. We discovered the "advantages" of what people in poor living conditions have over people who live in clean sterile environments.

We were pleased to have a reference base in the "hygiene hypothesis" [1] and we postulated that adverse conditions in early life could confer lifelong protection against G.I.T. diseases, even if there was transgression of the dietary principles regarded as good practice in HICs.

Other fact that surprised the GIT doctors was that colonoscopy always revealed a "Clean Colon" even in aged people. There were no diverticulae or polyps, and the colon was distensible. In contrast, the White population in Johannesburg showed a tight rigid colon, often with diverticulae and colonic polyps. To ensure that our Fellows from Baragwanath Hospital had experience in colon disease, we sent them to the Johannesburg Hospital to gain an all round experience.

The advantage of investigating the rarity of GIT diseases in urban Blacks was that we witnessed the emergence of HICs diseases in our midst. As city life took its toll on the Black population and GIT diseases, previously rare, began to emerge, we were in an advantageous situation to observe these changes and attempt to understand the reasons for this transition. This experience and its consequences for Sub-Saharan Africa form the core of this book.

REFERENCE

[1] Strachan DP. Family size, infection and atopy: the first decade of the "hygiene hypothesis". Thorax 2000;55(Suppl 1):S2–10.

CHAPTER

ALCOHOL AND SMOKING: IMPACT OF BEHAVIORAL RISK FACTORS

1

Isidor Segal

G.I.T. Unit, Chris Hani Baragwanath Hospital and University of Witwatersrand, Soweto, South Africa

INTRODUCTION

ALCOHOL CONSUMPTION

Global situation

Alcohol is one of the top three priority areas in public health. Although only half the global population drinks alcohol, it is the world's 3rd leading cause of ill health and premature death after low birth weight and unsafe sex [1].

Alcohol is a recognized risk factor for morbidity and mortality. It contributes to or is the sole cause of chronic health problems because of its direct toxic effect on organs as in alcoholic liver cirrhosis and pancreatitis. Its intoxicating properties result in accidents and injuries. It is a dependent producing substance and contributes to social problems including domestic violence, family disruption, and all types of intentional and unintentional injury, including homicides, suicides, violence in the street and workplace problems such as absenteeism, low productivity, and inappropriate behavior [1, 2]. The harm from alcohol disproportionately affects poorer people. In digestive diseases alcohol is a cause of liver, pancreatic, and gastrointestinal diseases. It also increases the risk of HIV/AIDS, tuberculosis and community-acquired pneumonia. The real absolute risk of dying from an adverse alcohol-related condition increases with the total amount of alcohol consumed over a lifetime.

ALCOHOL USE IN SSA

Alcohol has been an integral part of African social life for centuries. Except where it is banned for religious reasons, large quantities of brewed or distilled drinks are produced in local communities or by modern commercial enterprises to satisfy the tastes of a growing number of consumers. This includes even remote villages. Commercially produced beer is the most preferred drink and western spirits have usurped the cultural roles previously reserved for traditional drinks [2].

Data from the W.H.O. Global Alcoholic Database (GAD) [3] show a wide variation in per capita consumption of recorded alcohol in African countries ranging from less than 1L of pure alcohol in some (mainly Muslim) countries to more than 10L in a few of other countries. Today the average per capita consumption is a little under 4L, less than half the European average [2]. The figures, however, do not account for much of the alcohol consumed in the form of traditional beverages.

Digestive Diseases in Sub-Saharan Africa. https://doi.org/10.1016/B978-0-12-815677-3.00001-3

Consumption of commercial beverages is expected to rise in the coming years as the economic conditions continue to improve in some countries and as a result of increasing marketing and promotional activities by the industry.

Alcohol consumption and drinking patterns: Unrecorded alcohol consumption refers to consumption of beverages that are not part of official statistics on production and trade reported to the FAO. In this category are traditional beverages made from palm trees, a variety of grains and fruits, and drinks distilled from these local brews [2]. Traditional drinks contribute significantly to the overall consumption of alcohol in all African countries where drinking is a common practice. The overall estimate of unrecorded alcohol consumed in African countries is 50% [3].

In countries like Kenya, Rwanda, Seychelles, and Zimbabwe, unrecorded consumption accounts for much more of the total consumption of alcohol.

There is a high rate of abstention in every country—more than 50% of adults are abstainers, and among women it is often as high as 80%. In some countries, however, notably Uganda and some regions in Nigeria, male and female drinkers, respectively, engage in heavy episodic drinking. The difference between men and women disappears when the focus is on drinkers only. For example, in Nigeria, South Africa, and Ethiopia, more women drinkers than men reported regular consumption of volumes of alcohol that exceed what can be defined as moderate drinking [2].

Most drinking takes place in bars and at parties during weekends, indicating the social nature of drinking in African countries. An example of the characteristic progression of alcohol consumption as people migrate from a mainly rural to an urban society is characterized by what has happened in one of the economic powerhouses of Africa-South Africa.

Alcohol consumption patterns in South Africa: South African adult per capita alcohol consumption (APC) in 2005 equaled 9.5 L of pure alcohol. Of this consumption 26.3% was homemade and illegally produced alcohol. APC in South Africa is above the rest of the world average of 6.13 L, and the regional average for Africa of 6.2, but below the European regions APC of 12.2 [4]. South Africa's recorded APC showed an increasing trend until the mid-1990s, but over at least the last decade per capita demand has been shrinking—this is mainly due to the fact that South Africa is a saturated market, However those who do drink appear to do so at binging levels. 45.4% have weekly binges In South Africa, 65% of the population has never consumed alcohol—this is among the highest abstention rates in the world Of the 35% of the population that do consume alcohol, only a small percentage consume branded products. The largest percentage consume home brews or illegal products.

On analysis of pure alcohol consumed by the drinking population, South Africa has the 75th highest APC ratio in the world [4]. Consistent with global findings, alcoholism shows an increasing trend in lifetime prevalence use among youth aged 13–19 years in South Africa [5].

The small population of drinkers in Africa drinks excessively—i.e., heavy episode drinking. In South Africa, there are high levels of youth drinking of alcohol. In many African countries, there has been a change in the type of ingredients used in making beer with sorghum being replaced by maize.

Alcohol consumption is associated with cancer of the esophagus and the emergence of acute and chronic pancreatitis.

SMOKING

Active and passive smoking is a global problem and is a factor in the etiology of cancers, cardiovascular and respiratory diseases.

The majority of more than 1 billion smokers worldwide now live in low- and middle-income countries. An estimated 60% of men in some countries in Eastern Europe and East Asia smoke. The prevalence of smoking among women is still highest in HMIC societies. The prevalence of smoking remains relatively low in sub-Saharan Africa and smokers smoke fewer cigarettes than do their Western and Asian counterparts [6]. Significantly, the type of cigarettes smoked in many African countries is different to that of other regions of the world. In South Africa, a study of 200 esophageal cancer (EO) patients and 391 controls showed that 75% of patients smoked cigarettes compared to 46% controls [7]. The survey also included questions about the use of hand-rolled cigarettes in which pipe tobacco is used wrapped in newspaper, brown paper, or telephone directory paper. Such cigarettes were often preferred because of their flavor and their slow-burning quality. They were also cheaper than commercial cigarettes. Tobacco smoked either in pipes or hand-rolled cigarettes was found to be a major carcinogenic stimulus in black South African populations. The crude relative risk (RR) (RR: 1.07) indicated that there was no risk involved in smoking only commercial cigarettes. However, in conjunction with other predictors, it becomes an important risk factor for esophageal cancer. Risk increased with increasing consumption for the smokers of both hand-rolled and commercial cigarettes [7].

EPIDEMIOLOGY OF ESOPHAGEAL CANCER-RELATIONSHIP TO SMOKING, ALCOHOL, AND DIET [7,8]

Esophageal cancer (EO) was a rare disease in the South African Black population until the last few decades. Increased incidence has occurred and at present it is the commonest cancer in Black men in many parts of South Africa. Among the risk factors identified as agents in the etiology of the disease are tobacco and alcohol abuse. Moreover, consumption of traditional beer made from maize was found to be a major risk factor. This predisposition probably arises from a lifelong mild deficiency, usually subclinical, of several micronutrients. The association between smoking pipe tobacco and consumption of traditional beer was found to be a major risk factor in the etiology of esophageal cancer.

PREDICTIONS FOR THE 21ST CENTURY

The consumption of Western-type alcohol will increase in Africa mainly due to increased urbanization and exposure to promotion of alcoholic drinks by the media. However, traditional home brews will still play a prominent role in the drinking pattern because it is cheaper than Western-type spirits and also because it is part of the social fabric of the African way of life.

Smoking of commercial cigarettes will also increase. The reason being exposure to advertising from the tobacco manufacturers, the ease of availability, and urbanization accompanied by a more sophisticated life style.

While Africa is the poorest region in the world with the highest rates of poverty, it is not becoming poorer. In fact, it is one of the faster-growing regions in the world today. Africa is experiencing an economic boom. This has resulted in the growth of the number of smokers and cigarettes smoked in Africa. Africa represents the future epicenter of the tobacco epidemic [9].

As a result of successful tobacco control South Africa, which in 1990 accounted for 39% of the African market, in 2010 it accounted for only 17%. This was a result of the intervention by public health authorities [9]. While many African markets have low smoking prevalence, these countries will likely evolve to later stages of the epidemic with increased smoking prevalence. A large number of

African countries are in the early stages of the epidemic. By 2060, Africa will have the second-most smokers of any region, behind Asia with 14% of the world's smokers (from the current 6%) and by 2100 Africa will be home to 21% of the world's smokers [9].

In conclusion, the scenario for Africa in the 21st century will probably result in a significant increase in alcohol consumption and cigarette smoking. Probable reasons for this are upward mobility of the African population, and the promotion of Western-type alcohol and cigarettes by the media. However, traditional home brews will still play a prominent role in the drinking pattern because it is cheaper than Western-type spirits and also because it is part of the social fabric of the African way of life.

Public health authorities will have to counter the efforts of the commercial groups; otherwise, the consequences may be catastrophic.

REFERENCES

[1] Anderson P, Moller L, Galea G. Alcohol in the European Union. Consumption, harm and policy approaches. World Health Organisation; 2012. p. 1–32.

[2] Obote IS. Alcohol use and related problems in Sub-Saharan Africa. Afr J Drug Alcohol Stud 2006;5(1):18–26.

[3] World Health Organisation. Global status report on alcohol 2004. Geneva: WHO; 2004.

[4] World Health Organisation. Alcohol demand/consumption patterns in South Africa. World Health Organisation: Global Status Report on Alcohol and Health 2011, Geneva: World Health Organisation; 2011 (Country Profiles: South Africa).

[5] Seggie J. Alcohol and South Africa's youth. SAMJ 2012;1012(7):1–3.

[6] Ezzati M, Riboli E. Behavioural and dietary risk factors for non-communicable diseases. N Engl J Med 2013;369(10):954–64.

[7] Segal I, Reinach SG, De Beer M. Factors associated with oesophageal cancer in Soweto, South Africa. Br J Cancer 1988;58:68–686.

[8] Bradshaw E, McGlashan ND, Harrington JS. Oesophageal cancer: smoking and drinking in the Transkei. Occasional paper 27 Grahamstown: Institute of Social and Economic Research, Rhodes University; 1982 [abstract].

[9] Blecher E, Ross H. Tobacco use in Africa: tobacco control through prevention. American Cancer Society; 2013. p. 1–16.

CHAPTER

URBANIZATION, UPWARD MOBILITY, DEMOGRAPHICS, ENVIRONMENTAL CHANGES, AND AVAILABILITY OF NATURAL RESOURCES IN SUB-SAHARAN AFRICA (SSA)

2

Isidor Segal

G.I.T. Unit, Chris Hani Baragwanath Hospital and University of Witwatersrand, Soweto, South Africa

URBANIZATION

SSA consists of 46 different countries [1]. It is the least urbanized region in the world with only 39.1% of the region's population living in cities [2]. However, the urban population is projected to more than double to reach 1 billion by 2050 [2]. Urbanization is a health threat for certain vulnerable populations and this demographic shift threatens a humanitarian disaster [3].

Urbanization in Africa is linked to poverty since more than two-thirds of urban dwellers live in slums, noted to be the highest proportion in the world [3]. Rapid urbanization leaves cities unable to provide basic services [1], and as a consequence results in poor health in many slum dwellers [2]. In 2000, 30%–50% of African urban dwellers lacked a safe water supply and across Africa almost half of the urban population lacked access to improved sanitation [2]. Open defecation is the only sanitation practice available to a third of the population and contributes to the contamination of water, land, and many of the waterborne diseases that are prevalent in slums. These include malaria, dengue, and yellow fever with children bearing a disproportionate burden of disease in slums. In Ethiopia and Niger, the rate of child malnutrition in urban slums and rural areas is close to 40% and in Nairobi 60% of the city's population live in slums where child mortality is 2.5% greater than other city areas [2]. It is noteworthy that urban pollutants in Africa are responsible for an estimated 49,000 premature deaths annually.

A further factor affecting the populations is the lack of electricity. Two-thirds live without electricity accounting for half the people without electricity in the world [1].

An apt description is given by the United Nations Development Programme is "the urbanization of poverty" (Forbes and Lindfield) [4].

Digestive Diseases in Sub-Saharan Africa. https://doi.org/10.1016/B978-0-12-815677-3.00002-5

UPWARD MOBILITY

Upward mobility is defined as rising from a lower to a higher social class or status (www.dictionary.com/browse/upward mobility).

The impact of upward mobility in SSA is implicit in the urbanization process. Rural immigrants leave a rural tribal sociocultural milieu for the multiethnic Western environment of the city; they move from a subsistence economy into one requiring new skills, new concepts of time, new work attitudes, and different kinds of work relationships. They are removed from the close neighborliness of a traditional community to the anonymity of city life. Migrants are subjected to increased stress in coping with the "new environment." As certain individuals gain in skill and proficiency and exhibit certain skills and characteristics, they become upwardly mobile and move upwards out of their ranks [5]. Upward mobility has been associated with duodenal ulcer, cardiovascular disease, and psychiatric disturbances [6–9].

DEMOGRAPHY

Globally populations are aging, becoming more urbanized and more mobile. SSA is unusual in serving a growing population of young people, who are a source of competitive labor and a growing consumer market [10, 11]. This should be considered in a background where close to two-thirds of Africa's unemployed are youth. It is estimated that by 2050 almost 20% of the world's population will live in Africa, up from 7% in 1950 and this will comprise a young population. By 2050, young people 15–25 years of age will account for 1 person in 5 [3]. It is estimated that the African population will increase from 910.4 million to an additional 1.2 billion by 2050 [10].

ENVIRONMENTAL INFLUENCES AND IMPACTS

Xenobiotics
Smoking
Alcohol
Volatile Hydrocarbons
Occupations
Air Pollution
Dietary Changes
Climate Change

XENOBIOTICS

Xenobiotics are substances foreign to living systems and include drugs, pesticides, pollutants, carcinogens, volatile petrochemicals, food additives, and polluted working environments.

SMOKING

Smoking is a risk factor for many diseases, lung cancer being the foremost. In SSA, it is a leading risk factor in esophageal cancer [12].

Smoking has been identified as the single greatest preventable cause of death in the world today and kills 5.4 million people a year [13].

ALCOHOL

Alcohol leads to health-related and social problems. In the digestive system, alcohol is a leading cause of cirrhosis and pancreatitis and is also related to cancers of the mouth, oropharyngeal, esophagus, liver, colorectum and diabetes.

Also long-term alcohol abuse and smoking have been cited as a major cause of pancreatitis, which appears to be endemic among Sowetans, (Johannesburg) South Africa [14].

VOLATILE HYDROCARBONS

In SSA, exposure to volatile hydrocarbons and particularly petrochemicals increases susceptibility to pancreatitis and diabetes.

Chronic exposure to xenobiotics such as smoke from coal stoves and kerosene from Primus stoves are a feature of urbanized regions of SSA. This has been highlighted in a study from Soweto, Which Showed Increased Susceptibility to pancreatitis [14].

OCCUPATIONS

Xenobiotic exposure may derive from various occupations and have serious deleterious effects on health [14]. Braganza in 1986 suggested that occupational exposure to volatile hydrocarbons may be related to idiopathic and alcohol-related pancreatitis [15].

AIR POLLUTION

Air pollution is an important cause of increased morbidity and mortality worldwide. It has been suggested that sustained reductions in fine-particulate air pollution exposure should result in improved life expectancy [16].

A characteristic sight in SSA cities is the smoky haze that engulfs cities due to the fumes from coal stoves.

DIETARY CHANGES

SSA is experiencing a rapid nutrition transition (NT) with a resultant epidemiological transition, usually characterized by a decrease in undernutrition and infectious disease and an increase in overnutrition, overweight, obesity, and other noncommunicable diseases (NCDs) [17].

Dietary recommendations to prevent these unfortunate outcomes of the NT necessitate an understanding of the causes of the double burden, including interactions with poverty and economic development, as well as the biological (epigenetic) and sociological links between under- and overnutrition [17].

CLIMATE CHANGE

It is suggested that climate change may result in a increase of water- and food-borne diseases. Also rodent-borne and vector-borne diseases will increase—particularly malaria and diarrheal diseases.

An example of the influence of climate change is being experienced in the spread of malaria to the previous malaria-free region of the Eastern Highlands of Kenya [18, 19] Cholera risk may be amplified as an example of water-borne infection, mosquito-borne infections tend to increase with warming and changes in rainfall. Similarly, tick-borne infections may increase. It is evident that swift and aggressive international action is required to ameliorate the situation.

SUSTAINABILITY

SSA is making major strides in the economic, social, and political fronts. Since the turn of 21st Century, the estimated growth rate reached 3.9% in 2014. This has been assisted by strong commodity prices, new resource exploitation, and the improved domestic conditions contributing to several years of solid trend growth in the region's low-income countries [10].

The continental expansion of trade and investment is likely to advance due to the global appetite for natural resources. It offers an excellent opportunity for profitable new investments. Future global economic growth will increasingly come from emerging markets and SSA will be an important player. Before the global financial crisis, SSA had been growing rapidly with an average growth faster than Latin America, Europe, and Central Asia [9].

The potential for sustained growth will increase with modernization, recent sustained improved political and macroeconomic stability. There is political commitment to private sector growth and increased investment in infrastructure and education and throughout the African Continent, fiscal soundness and monetary discipline are increasing. Sovereign credit ratings in parts of the continent enjoy a positive outlook [9].

RESOURCES AVAILABILITY

SSA will outstrip Russia as a global gas supplier by 2040 according to the International Energy Association (IEA) spurring the European Union to invest in the region's development [10, 20]. It is emerging as a major new player in the natural gas market.

Since 1990 SSA has almost tripled its level of exports and diversified its trade and investment partners, e.g., during this time Chinese imports from SSA increased from $64 million to over $13 billion [10].

Natural resources will be a key source of export revenue with global demand increasing [10]. Today the region is not only a major supplier of natural resources but also the region with the greatest potential of new discoveries [11]. The World Bank predicts that both energy and food prices, driven predominately by the emerging economies resource needs, will remain high over the next 20 years [10]. The resource demand will also increasingly target renewable sources of energy and SSA is particularly well positioned for developing solar, hydro energy and the production of biofuels. While biofuels have already contributed to higher food crop prices, they also represent an opportunity for profitable production in the developing world [10]. Global markets for nontraditional exports such as horticulture and agriculture are expanding rapidly [10]. On the IT front, SSA has made great strides and is the fastest-growing region in the global cellular market. More than 4 million mobile phones are used daily [1]. SSA is faced with many challenges to realize such potential and will need to address the constraints to advance its competiveness. Opportunities will arise to improve the health care systems and management while business, governance, infrastructure, logistics, and education are implemented [8]. Critical policy areas will be adapted, reformed, and modernized to attract strategic investments in the underlying foundations that nourish business continuity [10].

The assessment of advances in digestive disease will include *treatment* such as Hepatitis C where several new classes of therapies are now available with short, convenient courses of interferon-free, fixed-dose single-pill regimens, with adverse-effect profiles that are markedly lower than those of past treatments [21], advances in *technology* such as diagnostic methodology, stem cell research, and particularly the progression of communication facilities will enhance health education. This will include improved workforce planning so that health professionals are encouraged to remain in their countries and not migrate to the "richer countries" [22].

REFERENCES

[1] UNDP. Africa at a glance, http://.africa.undp.org/content/rba/en/home/regioninfo.html.

[2] Ramin B. Slums, climate change and human health in sub-Saharan Africa. Bull WHO 2009;87:886. https://doi.org/10.2471/BLT.09.073445.

[3] Patel. Urbanisation an emerging human disaster. N Engl J Med 2009;361(8):741–3.

[4] Forbes D, Lindfield M. Urbanisation in Asia. Lessons learned and innovative responses. Australian Agency for International Development (AusAID); 1997. p. 1. Executive Summary.

[5] Segal I, Dubb AA, Ou Tim L. Duodenal ulcer and working class mobility in an African population in South Africa. Br Med J 1978;1:469–72.

[6] Levenstein S. "Stress and peptic ulcer" life beyond helicobacter. BMJ 1998;316(7130):538–41.

[7] Mbanya JCN, Motala AA, Sobngwi E. Diabetes in sub-Saharan Africa. Lancet 2010;375(9733):2254–66.

[8] Agyepong IA, Sewankambo N, Binagwaho A, et al. The path to longer and healthier lives for all Africans by 2030:the *Lancet* Commission on the future of health in sub-Saharan Africa. Lancet 2017;390(10114):2803–59.

[9] Chibanda D. The future of psychiatry in Africa—thinking outside the box. Lancet Psychiatry 2017;4(10):741–2.

[10] IMF. Regional economic outlook: sub-Saharan Africa. Sustaining growth amid global uncertainty. April 12 http://www.imf.org/external/pubs/ft/reo/2012/afr/eng/sreo0412.htm; 2012.

[11] Ngozi-Okonjo-Iwesia. Fulfilling the promise of sub-Saharan Africa, http://www.mckinsey.com/insights/economic studies/fulfilling he promise of sub-.

[12] Segal I, Renach SG, de Beer M. Factors associated with oesophageal cancer in Soweto, South Africa. Br J Cancer 1988;58:681–6.

[13] WHO Report on the Global Tobacco Epidemic. The MPOWER package, www.who.int/tobacco/mpower/en/; 2008.

[14] Braganza JM. A framework for the aetiogenesis of chronic pancreatitis. Digestion 1998;59(Suppl. 4):1–12.

[15] Braganza JM, Jolley JE, Lee WR. Occupational volatile chemicals and pancreatitis: a link. Int J Pancreatol 1986;1:9–19.

[16] Andreau K, Leroux M, Bouharrour A. Health and cellular impacts of air pollutants: from cytoprotection to cytotoxicity. Biochem Res Int 2012;2012:493894.

[17] Vorster HH, Kruger A, Margetts BM. The nutrition transition in Africa: can it be steered into a more positive direction? Nutrients 2011;3(4):429–41.

[18] United States Environmental Protection Agency. [Online] Available: www.epa.gov/climate change. Schuman EK, Global climate change and infectious diseases. N Engl J Med 2010;362(12):1061–3.

[19] McMichael AJ, Woodruff RE, Hales S. Climate change and human health: present and future risks. Lancet 2006;367:859–69.

[20] IEA: Sub-Saharan–Africa will produce more gas than Russia. EurActiv. http://www.euractiv.com/sections/energy/iea-sub-saharan-africa-will-produce-more-gas…
[21] Jayasekera CR, Barry M, Roberts LR, et al. Treating Hepatitis C in lower-income countries. N Engl J Med 2014;10.
[22] Hunter DJ, Fineberg HG. Convergence to common purpose in global health. N Engl J Med 2014;370:1753–5.

CHAPTER

REVIEW OF TRADITIONAL DIGESTIVE DISEASES COMMON IN SSA

3

Isidor Segal

G.I.T. Unit, Chris Hani Baragwanath Hospital and University of Witwatersrand, Soweto, South Africa

INTRODUCTION

A review of traditional digestive diseases that prevailed in SSA populations before the rapid rise in urbanization will be discussed:

CANCER OF THE ESOPHAGUS

Cancer of the esophagus was an uncommon disease in South African Blacks during the 1920s and 1930s [1]. In West Africa, it was practically unknown. The age-standardized incidence rates in Southern Africa were 19.7 (100,000), Eastern Africa 19.1, Central Africa 1.5, and Western Africa 1.3 per 100,000 [2].

Since the 1930s, an alarming increase in incidence occurred in South Africa and it is the commonest cancer in Black men. The cancer is of the squamous cell type. Adenocarcinoma is rare. The findings of a case-control study in Soweto support the hypothesis that the consumption of beer made from maize was a major factor in the development of esophageal cancer in Africa [1]. Cook [3] suggested that both the geographical and temporal distributions in Africa could reflect the use of maize as a major ingredient of beer.

An additional risk factor in Black South African populations was tobacco and especially pipe tobacco smoked in either pipes or in hand-rolled cigarettes [1].

The evidence suggests that any effect may occur through nutritional deficiencies that are engendered and through the promoting effects of alcohol rather than through the direct impact of a chemical carcinogen [4, 5].

GERD

Epidemiological and clinical studies that have reported on GERD, Barrett's esophagus, esophageal adenocarcinoma, and *Helicobacter pylori* infection in SSA were reviewed. The data indicate that Barrett's esophagus is rare, GERD and esophageal adenocarcinoma are uncommon in all regions of SSA studied (South Africa, Ethiopia, Nigeria, Zimbabwe, Kenya, and Uganda.) Hiatus hernia is also uncommon [6–8].

Digestive Diseases in Sub-Saharan Africa. https://doi.org/10.1016/B978-0-12-815677-3.00003-7

GASTRIC ULCER IN SSA

Gastric Ulcer (GU) was uncommon in the Black populations of SSA, the ratio of DU to GU being 3:1 [9]. The disease was also rare in other countries of SSA [10]. In 1927 Beyers noted that GU was very rare in Johannesburg [11], and Keeley [12] reported only 14 cases per year at Baragwanath Hospital for the period 1954–56.

The most common presentations of GU were dyspepsia and hemorrhage. The vast majority of the ulcers were located in the antrum (68%) [9].

Men and women were equally affected and the mean age of GU patients was 45.6 years. Fifty per cent of the patients regularly used analgesics [9].

GASTRIC CANCER (GC)

Gastric cancer was uncommon in SSA populations. Although there is a paucity of data, a study by Walker et al. [13] showed that the cancer mortality rate for GC in South African Blacks was 6/100,000 in men and 3/100,000 in women compared to the U.K. population, which showed 27/100,000 in men and 18/100,000 in women (1982) [13]. Higher rates would have been expected in urban dwellers. The low rates in large cities may well be valid for studies by social and other workers indicate that the huge majority of urban patients with cancer do ultimately seek medical attention [13].

DUODENAL ULCER

DUODENAL ULCER PREVALENCE IN AFRICA SOUTH OF THE SAHARA [14]

Duodenal ulcer prevalence is high in regions of high rainfall where the staple crops are yams, cassava, sweet potato, green bananas or where the staple diet is refined maize, wheat, or rice.

Duodenal ulcer prevalence is low in regions of lower rainfall where the staple crops are maize, millets, soya, or wheat. It was rare also in the lowlands of Ethiopia where maize, millet, and wheat were grown [14].

Type of ulcer: In the areas of high prevalence, the commonest complication was pyloric stenosis. Gastric ulcer was relatively rare [14].

DU IN AN URBAN MILIEU

There has been a steady increase of duodenal ulcer (DU) over the past 80 years in Soweto. DU is a "Western" type of DU with hemorrhage and perforation the major complications. This is different from the stenosing type of DU that occurs in rural areas of Africa [15].

ACUTE AND CHRONIC PANCREATITIS

Pancreatitis was previously an uncommon disease among Black populations in SSA. Pancreatitis in South Africa emerged as a common clinical problem 15–20 years after legislation prohibiting consumption of Western-type alcohol was repealed [16–18].

It is predicted that the pattern of pancreatitis in LIMCs will follow the trend demonstrated in South African Blacks when the consumption of Western-type alcohol becomes common in those countries.

TROPICAL PANCREATITIS

Tropical pancreatitis is a form of chronic pancreatitis (CP) characterized by recurrent abdominal pain, pancreatic calculi, and diabetes mellitus, occurring mostly among children and young adults of many Afro-Asian nations. Countries in which it occurs include Uganda, Nigeria, the Republic of Congo, Malawi, Zambia, Ghana, and Ivory Coast [19].

Etiology. The exact etiology of this disease has not been established. A number of recent observations doubt that protein malnutrition is the initiating factor of this disease. CP, alcoholic or tropical, has been hypothesized as one of the many diseases caused by unmitigated free radical (FR) injury. An association with a high consumption of diet rich in cassava root (tapioca, *Manihot esculenta*) and tropical pancreatitis was suggested in many previous reports. However, the association appears to be weak [20].

LARGE BOWEL DISEASE

APPENDICITIS

In low- and middle-income countries (LMICs), the frequency of appendicitis, although increasing slowly in some countries, e.g., Kenya is much lower than in high-income countries (HICs). In South Africa, among rural and urban Black adolescents, the frequency of appendectomy is only 5%–10% of that in local Whites of the same age [21]. The figures are low in Ghana, Madagascar, Central African Republic, Ethiopia, and Nigeria although increasing slowly, particularly in urban areas [22].

Segal et al. [21] reported that parasitic ova are very seldom observed (3.6%). Fecoliths were present in 20% of diseased appendices compared with White patients (50%–75%) [21]. There were about two cases of appendicitis per 1000 admissions of adults annually, a low frequency consistent with the tenfold differential of appendectomies reported to prevail in urban Black and White adolescents. The progressively changing diet of urban Blacks currently low in fat and dietary fiber has not yet become associated with a marked increase in the occurrence of the disease.

SIGMOID VOLVULUS

Sigmoid volvulus results from the rotation of the sigmoid colon on its mesenteric axis leading to a closed loop obstruction—it is a common cause of obstruction in SSA. Redundancy of the sigmoid colon with its meso-sigmoid parietal attachment can predispose the loop to twisting and volvulus. It affects males more than females and is associated with a high mortality [23].

Factors important in etiology are the consumption of a high fiber diet, which increases fecal bulk resulting from fermentation of carbohydrate in the colon and gas formation [23], and the drinking of traditional beer made from sorghum.

The incidence of sigmoid volvulus has been decreasing in the urban Black population of Soweto with urbanization and the increasing adoption of a Western type of refined food and alcohol made from refined maize [24].

DIVERTICULAR DISEASE (DD)

DD was virtually unknown in Africans living south of the Sahara and who have not changed their traditional high-residue diet [25–29]. As Africa becomes more urbanized and industrialized, DD becomes more evident [25]. Noteworthy is that the anatomic distribution of DD in Blacks showed that the major location of DD is in the descending colon, whereas in HICs the majority of diverticulae occur in the sigmoid colon [25].

The hypothesis suggested is that in Africans the high fiber diet, traditionally consumed in rural areas, has been replaced by a low-residue, fiber-deficient diet [25].

CONCLUSION

The urbanization of Africans and resultant change in diet has resulted in the emergence and increased prevalence of DD in SSA.

INFLAMMATORY BOWEL DISEASE (IBD)

The incidence and characteristics of IBD are similar in the White South African population compared to that of HICs populations. The Colored population (Caucasian, Malay, and Black) has a high incidence of aggressive, complicated IBD, in whom Cohn's Disease (CD) is more prevalent than Ulcerative Colitis (UC) [30, 31]. The incidence of IBD is significantly lower in the Black population of South Africa. UC in common with CD was rare in Black populations of SSA [32, 33].

It is proposed that in the context of poverty in infancy an adverse environmental situation conditions the gut and paradoxically, acts as a protective factor against subsequent digestive diseases in adulthood. It has also been proposed that the Hygiene hypothesis may be a factor in this low prevalence [34].

COLORECTAL CANCER (CRC)

There is a low incidence of CRC and colonic polyps among Black populations of SSA [35–37].

A study by Segal et al. in 1981 [35] showed that cancer of the large bowel is uncommon in South African Blacks, and that colorectal polyps do not appear to constitute a precursor to most colorectal cancers. Furthermore, it is shown that dietary factors associated with this cancer in High-Income Countries (HICs) populations are not evident in the Black population. In the South African White population, however, the disease behaves in a similar way to that observed in HICs. Other definitive differences found were the absence of multiple synchronous cancers and diverticular disease in the Blacks with colorectal cancer. It is thus postulated that dietary factors are absent, or have not been present for a sufficient length of time to influence the development of polyps or polyp-cancer sequence in this population. It is also possible that the adenoma-carcinoma progression observed in Western countries may not be relevant to the development of all colorectal carcinomas in communities such as those reported here.

BILIARY DISEASE

Gallstones and other biliary diseases were uncommon in SSA [38]. Although there are no accurate epidemiological data on the performance of cholecystectomies, it was not a common procedure in SSA. In 1927, Beyers [11] found that cholecystic disease was found in only four cases among 18,000 black patients at the Johannesburg General Hospital and again in 1934 Brebner [39] working at the same Johannesburg Hospital found that only 0.67 per 1000 patients were diagnosed with gallbladder disease. Biliary disease was noted to be uncommon in Nigeria [40].

LIVER DISEASE

VIRAL HEPATITIS [41]

Hepatitis A

The incidence of hepatitis A is strongly correlated with access to clean water and adequate sanitation and this factor has mostly contributed to the endemic nature of hepatitis A in SSA.

Hepatitis B (HBV)

HBV is highly endemic in SSA.

Hepatitis C (HCV)

The epidemiology of hepatitis C in SSA is incompletely understood with accurate epidemiological information very limited. A 2002 review of HCV epidemiology in SSA estimated the overall prevalence to be 3%, although some reports place the prevalence at 5.3%. Significant regional variation has been observed. A recent meta-analysis has demonstrated this wide regional variation in seroprevalence with 6.8% in the central Africa region, 4.3% in West Africa, and 0.9% in southeast Africa.

Hepatitis D (HDV)

Genotypes 5–8 occur in Africa, where HBV is endemic. Documented HDV seroprevalence rates vary geographically from low rates in countries south of the Equator: 0%–0.6% in South Africa, 0% in Mozambique, 0.6% in Kenya (south of Mount Kenya) to high rates north of the equator: 31% in Kenya (north of Mount Kenya), 7%–44% in Tunisia, 25%–28% in Sudan, 17%–56% in Somalia, 3% in Senegal, 6%–12% in Nigeria, 15%–33% in Mauritania, 9%–67% in Gabon, 5%–58% in Egypt, 2%–22% in Djibouti, and 6%–27% in Cameroon. Epidemic outbreaks have been reported in Gabon, Cameroon, Nigeria, and Mauritania.

Hepatitis E

In Africa, most outbreaks are sporadic with seroprevalence data from Zambia suggesting an overall seroprevalence of 42%; 28% in HIV-negative and 71% in HIV-positive adults. A recent study in the Western Cape in South Africa in both adults and children found an overall seroprevalence of approximately 30%.

HEPATOCELLULAR CARCINOMA IN SSA

Hepatocellular carcinoma is the second most common cause of cancer in men in SSA. Viral hepatitis is the most common risk factor associated with the development of this malignancy. It is likely to be significantly underestimated, particularly as a tissue diagnosis of HCC is rarely performed on the African subcontinent. The reports of hospital-based series of patients have been and remain biased by the local experience. Registry data are inaccurate and often not available at all. An age-standardized ratio of greater than 15 per 100,000 of the population per year is regarded as high incidence of HCC. The age-standardized incidence rate in SSA is 41.2/100,000 persons/year: for South Africa it is 16.0 for males and 4.5 for females, for Zimbabwe 30.7 for men and 17.1 for females, and in Swaziland 22.0 in males and 5.2 in females. The usual age distribution in HICs is late fifties to late seventies. In SSA, there is a distinct shift toward a younger age in Black males with a mean age of presentation around 45 years. Men have a higher incidence of HCC in SSA than do women in all populations. The higher incidence of HCC in this population probably reflects the higher rate of chronic HBV (approximately twice those in females). Dietary intake of carcinogens may be greater by virtue of the volume of food ingested and there may be differences in the efficiency with which ingested carcinogens are metabolized.

COMMENTS

Many digestive diseases common in HICs are uncommon in LMICs of SSA. These include GERD, esophageal adenocarcinoma, gastric cancer, small intestinal diseases, IBD, diverticular disease, and colorectal cancer.

Traditional diseases common in SSA were esophageal squamous cancer, sigmoid volvulus, viral hepatitis, and hepatocellular cancer.

Accompanying urbanization, however, a spectrum of diseases common in HICs are beginning to emerge in LICs.

REFERENCES

[1] Segal I, Reinach SG, de Beer M. Factors associated with oesophageal cancer in Soweto, South Africa. Br J Cancer 1988;58:681–6.

[2] Fock KM, Ang TL. Cancer of the oesophagus: intercontinental variations. In: Segal I, Pitchumoni CS, Sung J, editors. Gastroenterology and hepatology manual a clinicians guide to a global phenomenon. Australia: McGraw Hill; 2011. p. 207–16 [chapter 20].

[3] Cook P. Cancer of the oesophagus in Africa. Br J Cancer 1971;25(4):853–80.

[4] Bradshaw E, Schonland M. Smoking, drinking and oesophageal cancer in African males of Johannesburg, South Africa. Br J Cancer 1974;30:157.

[5] Van Rensburg SJ. Epidemiology and dietary evidence for specific nutritional predisposition to oesophageal cancer. J Natl Cancer Inst 1981;67:243.

[6] Sharma P, Wani S, Romero I, et al. Racial and geographic issues in gastroesophageal reflux disease. Am J Gastroenterol 2008;103(11):2669–80.

[7] Segal I, Solomon A, Ou Tim L, et al. Hiatus hernia in Johannesburg blacks. S Afr Med J 1980;58:404–5. Sept 6.

[8] Segal I. The gastro-oesophageal reflux disease complex in sub-Saharan Africa. Eur J Cancer Prev 2001;10(3):2009–12; june.

[9] Segal I, Noormohamed AM, Ranchod S, et al. Duodenal and gastric ulcer in Soweto. S Afr Med J 1983;64:777–8.

[10] Kark AE. The incidence and pattern of peptic ulcer in Indians and Africans in Durban. Gut 1961;161(2):363–9.

[11] Beyers CE. Incidence of surgical diseases among the Bantu races of South Africa. J Med Assoc S Afr 1927;1:606–12.
[12] Keeley KJ. Alimentary disease in the Bantu: a review. Med Proc 1958;4:281–6.
[13] Walker ARP, Odendaal MI, Segal I. Cancer patterns in different ethnic groups in South Africa. In: Vahouny GV, Kritchevsky D, editors. Dietary Fiber. New York. Plenum Press; 1986. p. 515 22.
[14] Tovey FI. Duodenal ulcer in Black populations in Africa South of the Sahara. Gut 1975;16:564–76.
[15] Segal I, Ally R, Mitchell H. *Helicobacter pylori*—an African perspective. Q J Med 2001;94:561–5.
[16] Segal I. Pancreatitis in Soweto, South Africa. Digestion 1998;59(Suppl 4):25–35.
[17] Segal I, Lerios M, MacPhail AP, et al. The genesis of chronic pancreatitis in the South African Black population. S Afr Med J 1988;74:384–6.
[18] Segal I, Lerios M, Grieve T. The emergence of chronic calcifying pancreatitis in a developing country. In: Gyr K, Singer MV, Sarles H, editors. Pancreatitis: concepts and classification. Amsterdam: Excerpt Medical; 1984. p. 417–20.
[19] Balakrishnan V, Nair P, Radhakrishnan L, Narayanan VP. Tropical pancreatitis-a distinct entity, or merely a type of chronic pancreatitis? Indian J Gastroenterol 2006;(25): 74–80.
[20] Pitchumoni CS. Chronic pancreatitis. In: Segal I, Pitchumoni CS, Sung J, editors. Gastroenterology and hepatology manual: a clinicians guide to a global phenomenon. Australia: McGraw Hill; 2011. p. 316–27 [chapter 30].
[21] Segal I, Paterson A, Walker ARP. Characteristics and occurrence of appendicitis in the Black population in Johannesburg, South Africa. J Clin Gastroenterol 1986;8(5):530–3.
[22] Saad AA, Makama JG, Mohammed U, et al. Epidemiology of appendicitis in Northern Nigeria: a 10 year review; Sub-Saharan African. J Med 2014;1(4):1185–90.
[23] Mehari H. Management of sigmoid volvulus in Eritrea, www.unige.ch/cyberdocuments/theses2002/MehariH/these_body.html.
[24] Segal I. Decreasing incidence of sigmoid volvulus at Baragwanath Hospital. S Afr Med J 1985;67:443.
[25] Segal I, Solomon A, Hunt J. Emergence of diverticular disease in the urban South African Black. Gastroenterology 1977;72:215–9.
[26] Vally M, Koto MZ, Govender M. An investigation of diverticular disease among black patients at George Mukhari Academic Hospital, Pretoria, South Africa. S Afr Med J 2017;107(2):137–9.
[27] Oluyemi A, Odeghe E. Diverticular disease at colonoscopy in Lagos State, Nigeria. Niger Med J 2016;57(2):110–3.
[28] Kiguli-Malwadde E, Kaszi H. Diverticular disease of the colon in Kampala, Uganda. Afr Med Sci 2002;2(1):29–32.
[29] Ihekwaba FN. Diverticular disease of the colon in Black Africa. J R Coll Surg Edinb 1992;3:107–9.
[30] Chu KM, Watermeyer G, Shelly L, et al. Childhood helminth disorder is protective against inflammatory bowel disease: a case control study in South Africa. Inflamm Bowel Dis 2013;19:614–20.
[31] Basson A, Swart R, Jordaan E, et al. The association between race and Crohn's disease phenotype in the Western Cape population of South Africa, defined by the Montreal classification system. PLoS One 2014;9(8):e104859.
[32] Segal I. Intestinal tuberculosis, Crohn's disease and ulcerative colitis in an urban Black population. S Afr Med J 1984;65:17–44.
[33] Segal I. Ulcerative colitis in a developing country of Africa: the Baragwanath experience of the first 46 patients. Int J Color Dis 1988;3(4):222–5.
[34] Segal I, Walker ARP. Persistent low prevalence of western digestive diseases in Africa: confounding aetiological factors. Gut 2001;48:730–2.
[35] Segal I, Cooke SAR, Hamilton DG, et al. Polyps and Colo-rectal cancer in South African Blacks. Gut 1981;22:653–7.

[36] Segal I, Edwards C, Walker AR. Continuing low colon cancer incidence in African populations. Am J Gastroenterol 2000;95(4):859–60.
[37] Irabor DO. Colorectal cancer: why is there a lower incidence in Nigerians when compared to Caucasians? J Cancer Epid 2011;2011:1–3. Article ID 675154.
[38] Walker AR, Segal I, et al. Prevalence of gallstones in elderly black woman in Soweto, Johannesburg, as assessed by ultrasound. Am J Gastroenterol 1989;84:1383–5.
[39] Brebner IW. Surgical conditions the liver including the gallbladder. S Afr Med J 1934;8:867–71.
[40] Njeze GE. Gallstones. Nig J Surg 2013;19(2):49–55.
[41] Sonderup M, Spearman W. Aspects of viral hepatitis in sub-Saharan Africa in the 21st century. See Chapter 15 in this book.

FURTHER READING

[42] Setshedi M., Davisson M. Hepatocellular cancer in Africa. See Chapter 16 in this book.

CHAPTER

EPIDEMIOLOGY OF DIGESTIVE DISEASES IN SSA: TRENDS THAT ARE OCCURRING

4

Isidor Segal

G.I.T. Unit, Chris Hani Baragwanath Hospital and University of Witwatersrand, Soweto, South Africa

CANCER OF THE ESOPHAGUS

Globally, IARC reported in 2002 that cancer of the esophagus is the 8th commonest cancer and the 6th commonest cause of cancer deaths worldwide, while in Sub-Saharan Africa it is the 4th commonest cancer(6%) [1, 2]. The cancer is on the increase in SSA with uneven geographical distribution. It is the leading cancer in Southern and East Africa and rare in West Africa. There are certain high-risk areas in Kenya [2].

During preindependence of African countries, there were three marked peculiarities of esophageal cancer reported, namely, geographical distribution, vagaries of the sex distribution, and changing pattern of incidence with time [2]. It is noted that most of the data are incomplete and anecdotal in many of the SSA countries. In 2002, it was only South Africa and Mauritius, which submitted the data of cancer mortality to WHO. The highest prevalence is in South and East Africa with squamous cancer the dominant cell type compared with adenocarcinoma.

In Ghana, the national trend of EC shows increasing incidence over the past 2 decades with alcohol and smoking the major risk factors (82% drank alcohol and smoked) [3]. As in other SSA countries, squamous cell cancer is the dominant cell type (78% vs. adenoncarcinoma (21.3%)). In Kenya, 98% of the cancers were of the squamous cell type and as in other SSA countries late presentation was usual.

There are many reports from South Africa about EC. It was a rare disease in the South African Black population until the last few decades. Increases in incidence have occurred and at present it is the commonest cancer in Black men in many parts of South Africa. Studies of western populations have indicated a multiplicative effect of alcohol and tobacco in the development of EC with risk increasing more sharply with rising tobacco consumption [4].

The rapid urbanization of SSA has resulted in an increase in consumption of spirits and smoking of commercial cigarettes in place of traditional beers and hand-rolled cigarettes. It is possible that adenocarcinoma of the esophagus will as a result increase with a commensurate decrease in squamous cancer.

Digestive Diseases in Sub-Saharan Africa. https://doi.org/10.1016/B978-0-12-815677-3.00004-9

GERD [5]

There are limited number of studies that have reported on GERD in SSA. There are no population-based studies describing the prevalence of GERD in SSA. The data indicate that GERD, Hiatus hernia, and Barrett's esophagus are rare in SSA (South Africa, Ethiopia, Nigeria, Zimbabwe, Kenya, and Uganda). Clearly, there is a recognizable potential for reporting bias based on differentials as compared to other regions of the world. This would include access to care, patient expectations, and willingness to undergo treatment, local resources, regional variations in diagnostic approaches, among other potential variables.

A study of hiatus hernia and reflux symptoms from 1970 to 1972 reported the frequency of reflux symptoms in only 23 subjects (2.2%)and hiatus hernia in only 4 cases (0.39%) [6, 7].

It may be anticipated that with upward mobility toward a modern life style the GERD complex should increase. However, the dilemma of *Helicobacter pylori* and its role in the protective effect against GERD will require further research to establish a positive direction.

GASTRIC ULCER (GU)

Gastric Ulcer (GU) was uncommon in the populations of SSA, the ratio of DU to GU being 3:1. In rural Africa, the ratio was 33:1 [8].

There are only a few reports of GU in SSA and there is considerable variation in the DU:GU ratios ranging from 0.25:1 in Ibadan Nigeria [9] to 8.2:1 in Nigeria [10]. A recent study from Nigeria [11] indicated that the prevalence had increased, while DU had declined. The use of NSAIDS (nonsteroidal anti-inflammatory drugs) was thought to be a relevant factor. Ameh [12] noted that in Zaria (Nigeria) in 1998 the ratio between perforated duodenal ulcer and perforated gastric ulcer had more than doubled from 2.6:1 to 6.5: 1.

It is noteworthy that a retrospective endoscopic review of dyspeptic patients from 12 African countries by Kidd et al. reported 26.4% of patients to be diagnosed with DU while 7% were diagnosed with GU. Of these, *H. pylori* was detected in 75% [13].

Numerous studies have shown a relationship between NSAIDS and GU [14, 15]. Urbanization of SSA populations and increased spending power will inevitably lead to an increased use of NSAIDS with a concomitant increase of GU.

GASTRIC CANCER

Based on Globocan Fact Sheets, almost one million new cases of stomach cancer were estimated to have occurred in 2012 making it the fifth most common malignancy in the world (globocan.iarc.fr/old/Fact Sheets/cancers/stomach-new.asp).

Data from the report in 2002 [16] show a total of 13,800 of stomach cancer in men and 10,700 in women in SSA in 2002. Age standardized incidence rates in males varied, per 100,000, from 3.4 in eastern, 8.2 in southern, and 13.4 in central Africa. In western Africa, where the incidence of stomach cancer is lowest, the male to female ratio is 0.9–1; however, there is a male preponderance in other areas. Although there is a generally low incidence rate in Africa, some populations have a particularly high incidence rate. This applies to the Colored population of South Africa in whom the rate is 98 per 100,000 and people living in the Great Lakes region [16]. There is a slight but not significant increase in incidence over time in Kampala, and reported increased incidence in rural Kenya (an endoscope was acquired). However, there was no change in incidence in South Africa between 1948 and 1964 (1.5%) [16].

In a study to determine if pathological changes accompanying *H. pylori* infection might explain the varying prevalence of stomach cancer in Uganda, Wabinga [17] investigated three ethnic groups resident in Uganda two of which have a high prevalence GC (Nyarwanda and Nkole) and Uganda, which has a low prevalence GC. This showed that in the two high GC-risk regions (Nkole and Nyarwanda) gastric atrophy was common, with *H. pylori* being detected in more severe forms of GA. Further, in the two high-risk regions, intestinal metaplasia (IM) was also observed, with a number of these cases showing the presence of *H. pylori*. Based on these findings, Wabinga concluded that the major histological features relating stomach cancer to *H. pylori* were presence of the infection in IM and GA, the two high GC areas (Nyarwanda and Nkole).

The question of *H. pylori* and its relationship to GC has not yet been resolved. Kidd et al. [13] concluded that it was difficult to establish a correlation between *H. pylori* and metaplasia and GC due the lack of good epidemiological data.

Evidence to support the African enigma has been reported by Segal et al. [18], who described the situation in Soweto, South Africa, where despite there being an extremely high prevalence of *H. pylori* (>90%), the incidence of GC is extremely low. Indeed based on a review of the Chris Hani Baragwanath, Soweto, hospital records, GC accounted for only 2.2% of all cancers diagnosed in the period 1948–64 and in 1992 the figure was 2.8% representing an average of 40 cases annually. This is extremely low given that the population of Soweto is over four million.

The prevalence of *H. pylori* in healthy adults and children remains high in the majority of people in SSA. In 1994, the IARC classified *H. pylori* as a carcinogen. This applies to HICs but does not correlate with the prevalence of GC in SSA.

Besides *H. pylori* other risk factors include alcohol, smoking, and dietary factors. The latter are rapidly changing and may herald an increased prevalence of GC in SSA.

DUODENAL ULCER (DU)

Tovey [19] has shown that DU is high in high-rainfall areas and low in low-rainfall areas of SSA. The major complication of DU in these regions is pyloric stenosis. In an urban milieu, the major complications of DU are perforation and hemorrhage, i.e., similar to that occurring in HICs.

High incidence rates have been reported in a number of major cities of Africa (Johannesburg, Durban, Nairobi, and Mombasa) [20]. In a study by Kidd et al., 26% of patients with dyspepsia had DU, and of these *H. pylori* was present in 90% [13].

The epidemiology of peptic ulcer disease, particularly in cities of SSA, is a reflection not only of the prevalence of *H. pylori* but also environmental factors including stress, which may function as a cofactor with *H. pylori* [21].

It is likely that DU and its complications will increase in prevalence in SSA as more people move from rural areas to cities. Because of poverty and lack of medical facilities, treatment of dyspepsia will be inadequate resulting in major complications of DU, i.e., perforation and hemorrhage presenting for emergency treatment to hospitals. Ameh [12] in a study from Nigeria showed an increased frequency of perforation in both DU and GU and that the ratio of perforated DU:GU more than doubled in the 1-year study period.

Medical facilities, endoscopic availability, and the ability to afford treatment with drugs such as proton pump inhibitors and therapy for *H. pylori* eradication (when present) will be necessary to improve the scenario.

PANCREATITIS

The types of pancreatitis occurring in SSA are alcoholic, biliary, and tropical pancreatitis.

ALCOHOLIC PANCREATITIS

Urbanization of SSA populations has been accompanied by increased alcohol and tobacco consumption and adverse changes in dietary patterns. As a consequence of these changes, alcohol-induced pancreatitis has become a common clinical problem in certain areas of SSA [22, 23].

ACUTE ALCOHOL-INDUCED PANCREATITIS (AAIP)

In African Blacks, AAIP is a severe disease associated with a high mortality and significant long-term morbidity in surviving patients [22, 23].

A 1997 study [22] of 136 patients showed male: female ratio of 3.8:1. *Alcohol was the predominately etiological factor in 83.1% biliary disease in 7.4% and idiopathic in 6.6%.* Substantial morbidity was encountered in 32.3%. 10.3% developed further pancreatic pathology such as pseudocysts, necroses, or an abscess. Mortality rate was 8.1%. Follow up after 9.3 months revealed serious morbidity in two-thirds of patients. 52% showed severe abdominal pain, 36% weight loss, and 18% a sonographically abnormal pancreas.

CHRONIC CALCIFIC PANCREATITIS [24–26]

Increased admissions for pancreatitis occurred some 20 years after Africans had access to Western-type alcohol. Case-control studies showed that approximately two-thirds still drank mainly traditional home-brewed beer of a low-alcohol content. 25 patients with chronic calcific pancreatitis were diagnosed in 1977. In the 3 year period, 1981–83, there were 55 new cases. A further survey in 1988 identified 90 new cases. Male-to-female ratio was 6:1. The mean age of the patients was 40 years and the mean alcohol intake was *150 g/day for an average of 15 years.*

It is significant that 20 years after Western-type alcohol was made available, chronic calcific pancreatitis (CCP) began to be diagnosed.

Case-control studies identified the same three environmental factors in each disease, namely, heavy alcohol consumption, marked exposure to occupational chemicals, and low intake of fruit (a major source of vitamin C).

It is predicted that the pattern of CCP in LIMCs will follow the trend demonstrated in South African Blacks when the consumption of Western-type alcohol becomes common in those countries.

BILIARY PANCREATITIS

The incidence in HICs of Acute pancreatitis varies from 1.7 (37 gallstones: 22 alcohol) in Wales, to 3 and 15 reported in most English studies. Other studies have found more equal cases of gallstone and alcohol, e.g., 30:29 in the north west of the Thames region [27].

Recently, there has been an increased incidence of biliary pancreatitis in SSA. This is related to the increased prevalence of gallstones and also to the HIV epidemic and treatment with the HAART regime.

TROPICAL PANCREATITIS

Tropical pancreatitis is a form of chronic pancreatitis (CP) characterized by recurrent abdominal pain, pancreatic calculi, and diabetes mellitus, occurring mostly among children and young adults of many Afro-Asian nations. Countries in SSA in which it occurs include Uganda, Nigeria, the Republic of Congo, Malawi, Zambia, Ghana, and Ivory Coast.

An association with a high consumption of diet rich in cassava root (tapioca, *Manihot esculenta*) and tropical pancreatitis was suggested in many previous reports. However, the association appears to be weak.

Data as to the incidence and prevalence in SSA of tropical pancreatitis are not known. It is noteworthy that in South India there is a high prevalence-114–200/100,000 population [28].

LARGE BOWEL DISEASE

APPENDICITIS

In 1986 [29], frequency of appendicitis was low in LMICs although increasing slowly in some countries (e.g., Kenya). Prevalence was much lower than in HICs. In South Africa, among rural and urban Black adolescents, the frequency of appendectomy was only 5%–10% of that in local Whites of the same age [29]. The figures were low in Ghana, Madagascar, Central African Republic, Ethiopia, and Nigeria although increasing slowly, particularly in urban areas [30].

There were about two cases of appendicitis per 1000 admissions of adults annually, a low frequency consistent with the tenfold differential of appendectomies reported to prevail between urban Black and White adolescents. The progressively changing diet of urban Blacks, low in fat and dietary fiber was not associated with a marked increase in the occurrence of the disease.

A later review of acute appendicitis in South Africa (2015) [31] showed that the trends in South Africa were changing and were consistent with what was happening in other LMICs. It was also noted that incidence estimates increased in South Africa from 8.2 cases per 100,000 population in 1986 to 15 per 100,000 over 20 years later [31].

Because of delayed presentation, South African patients presented with higher perforation rates. The perforation rate for appendicectomy patients was 36% and mortality rate was 1% [31]. Yang et al. [31] added that further research is needed to investigate specific factors, which delay care outcomes and cost analysis for laparoscopy surgery and the system strengthening of surgical services at district hospitals [31].

The increasing incidence of appendicititis is noted in other regions of Africa. In Nigeria, fear of surgery and cost of treatment were given as the main reasons for late presentations [32] and high perforation rates.

Ferris et al. [33] conclude that the appendicitis rate is stable in most HICs countries. Data from LMICs are sparse but suggest that appendicitis is rising rapidly.

INFLAMMATORY BOWEL DISEASE

IBD is uncommon in Black populations of SSA but appears to be increasing. This applies particularly to UC only 18 cases of UC and 7 CD in Black patients were reported prior to 1975 [34]. In a report on the global incidence of CD, the incidence in South Africa is noted as 1–3 per 100,000 people [35]. In other countries of SSA, the uncommoness of the diseases has been confirmed. In Nigeria, 3 cases of UC and 1 of CD were reported in 2011 in a 3-year period [36]. There have also been case reports of IBD from Sudan, Cote d'Ivoire, Senegal, Ethiopia, and Zimbabwe. However, recent reports from South Africa show that the number of new IBD cases (*all races*) diagnosed annually from 1970 to 2009 has risen from 12 patients to 87 patients [37, 38].

The incidence and characteristics of IBD are similar in the White South African population to that of HICs populations.

The Colored (mixed race) population has a high incidence of aggressive, complicated IBD, and in whom CD is more prevalent than UC. It is in the Colored population that a significant rise in IBD has occurred. In 1984, the incidence of CD in greater Cape Town region of the Colored, White, and Black populations was 1.8, 2.6, and 0.3/100,00 per year, respectively, and the incidences of UC 1.9, 5.0, and 0.6 per year, respectively [39]. In a report of a cross-sectional study of all consecutive CD patients seen at 2 large IBD referral centers in the Western Cape, South Africa, between September 2011 and January 2013, 194 CD subjects were identified: 152 (78%) were Colored; 35 (18%) White; and 7(4%) were Black [40]. A report by Deetlefs et al. of TB and an IBD cohort (615 patients) showed 394 (64%) were Colored, 149 (24%) White, 31 (5%) Black, and 41 (7%) Asian or unknown [41]. Coloreds formed the largest IBD demographic at 394 patients (64%), Whites 149 (24%), Blacks 31 (5%). 326 (53%) had CD and 270 (44%) had UC. There were 59 Coloreds who were TB positive (82%). Caucasian TB +ve 7 (10%); Black TB +ve 3 (4%). Race was the only risk factor identified for development of TB in the IBD cohort. There was a significant difference (*P*: .023) in complicated CD between White (9%) and Colored (60%).Coloreds are the largest IBD demographic and develop complicated CD and live in areas of high TB prevalence and more likely to develop TB. Therefore, Coloreds have aggressive CD with high rates of TB among IBD patients [38].

DIVERTICULAR DISEASE (DD)

DD is an age-related disease affecting approximately 25% of the population of HICs with prevalences ranging from 5% in individuals aged <40 years to 65% in those aged >65 years. Approximately only 10%–25% will manifest any related clinical symptoms. As many as 2/3 of individuals in the west have DD by the age of 85 [42, 43].

DD was virtually unknown in Africans living south of the Sahara and who have not changed their traditional high-residue diet [44].

In 1958, Keely reviewed hospital admissions at Chris Hani Baragwanath Hospital (Soweto)—the largest hospital in the African Continent with 3600 beds, and found no cases of DD. At the same time, Higginson and Simson found diverticulae in only one of 2367 necropsies of whom 789 were over 45 years of age. The paucity of the disease was confirmed in other studies from Johannesburg. This contrasts with a sample of the South African White population where in a series of 221 barium enemas 20.8% revealed diverticulae. Kloppers stated that one-third of middle-aged Whites in South Africa have diverticular disease [44].

As Africa becomes more urbanized and industrialized, DD becomes more evident. Urbanization includes changes in diet. In Africans, the high-fiber diet that was consumed in rural areas is replaced by a low-residue fiber-deficient diet [44].

In 1977, Segal et al. [44] reported 16 Africans with DD. The mean age was 48.9 years. All were urbanized people in whom a very high-carbohydrate intake with a very high refined carbohydrate intake, with a low fiber content was the rule. Most had been eating this refined food—low-fiber diet since childhood [44]. The series also showed a relative rarity of diverticula in the sigmoid colon. The reason for this is unknown. Thus, the urbanization of South African Blacks has been accompanied by the emergence of DD, a condition hitherto virtually nonexistent in this population [44].

A study in 2015, from Pretoria, South Africa [45], confirmed DD in the Black community—they found of 348 patients who underwent colonoscopies, 47 were diagnosed with DD—a prevalence of 13.5%.(confidence interval: 10.30–17.50) (95%) Mean age of the patients was 67 years. Lower gastrointestinal bleeding was the most common indication for colonoscopy. The left colon was the commonest site of involvement. The authors conclude that that there has been a considerable increase in the prevalence of DD among Black South Africans possibly owing to dietary changes and socioeconomic status [45]. In 2016, 7756 colonoscopies were performed at Baragwanath Hospital and 56 cases of DD were identified. Mean age of the patients was 62 (R. Ally, Personal Communication).

There are reports from other countries in SSA including Nigeria [46], Uganda [47], and Ghana [48].

CONCLUSIONS

DD is the commonest colonic disease in HICs populations but was rare in SSA. Following changes in dietary patterns DD has emerged and is becoming more prevalent in Africans in SSA.

SIGMOID VOLVULUS

Sigmoid volvulus is a common cause of obstruction worldwide. In the "volvulus belt" of Africa and the Middle East, 20%–50% of bowel obstructions are a result of volvulus almost exclusively of the sigmoid colon. Males are affected more than females (3:1 ratio) and it is associated with a high mortality [49, 50].

Factors important in etiology are (1) the consumption of a high-fiber diet, which increases fecal bulk resulting from fermentation of carbohydrate in the colon and gas formation [49], and (2) the drinking of traditional beer made from sorghum.

The incidence of sigmoid volvulus has been decreasing in the urban Black population of Soweto with urbanization and the increasing adoption of a Western type of refined food and alcohol made from refined maize [51].

It is likely that the occurrence of volvulus in SSA will follow the same pattern occurring in South Africa, i.e., a falling incidence of the disease concomitant with urbanization.

COLORECTAL CANCER (CRC)

Colorectal carcinoma (CRC) is the third most common cancer and is responsible for 1.2 million new cases per year worldwide and over 600,000 deaths annually [52].

Although previously colorectal cancer was considered rare in Africa, it is increasingly being diagnosed [53–59]. African patients present at a younger age as compared to their Caucasian counterparts and the disease tends to be more advanced. Patients <40 years have a similar clinicopathologic pattern to that of older patients. Atiological factors are similar to those seen worldwide. Disease profile mirrors international norms and is similar for young and older patients. Resection rate tends to be lower for Africans and for younger patients [56]. Incidence rates of CRC are higher in North Africa than in sub-Saharan Africa [55] and are nearly fourfold higher in HIC countries where it accounts for 9.4% compared with LMICs where it accounts for 2.5% of all malignancies [55–57]. The age standardized incidence is 21–45.7 per 100,000 in HICs and ranges from 3.3 to 7.2 per 100,000 in Africa [52].

COMMENTS

Concerns do exist, however, regarding the systematic under-reporting of CRC in African countries. SSA lags behind in terms of health facilities and research in colorectal cancer and other cancers but resources remain a challenge in populations of SSA [60–66].

BILIARY DISEASES IN SSA

The limited resources and poor socioeconomic environments in many SSA countries make this group of diseases extremely challenging to investigate and treat. In order to achieve high standards, centers of excellence are necessary.

The wide spectrum of disease burden ranging from parasitic infections, HIV (rare in HICs) to the more common diseases that plague HICs (i.e., gall stone disease) are prevalent within SSA populations.

GALLSTONES (CHOLECYSTOLITHIASIS)

Historically, gallstones were rare in SSA. Although there are no accurate epidemiological data on the performance of cholecystectomies, it was not a common procedure in SSA. In 1927, Beyers [67] found that cholecystic disease was found in only four cases among 18,000 black patients at the Johannesburg General Hospital and again in 1934 Brebner [68] working at the same Johannesburg Hospital found that only 0.67 per 1000 patients were diagnosed with gallbladder disease. Walker and Segal et al. reported on a series of 100 patients in 1985 who were screened for gallstones and 10% of women in the study were found to have stones. They suggested that an increase in incidence could be expected in the future [69].

However, over the last two decades, the number of cholecystectomies performed has increased dramatically and it now probably represents one of the more common operations carried out at Chris Hani Baragwanath Academic Hospital (CHBAH),where in excess of 300 cholecystectomies per year are done mostly laparoscopically. This is compared to 118 over 3 years as reported by Parek in 1987 [70].

The precise reason/s for this increase is not well defined. It is thought to be due to urbanization and an increased intake of a western-type diet. Walker et al. suggested that an increase in incidence could be expected in the future [69]. In our experience, this has occurred and we have noted an increase in incidence of cholecystolithiasis in both females and males. Pure pigment stones are uncommon.

GALLSTONE PANCREATITIS

There has been an increased prevalence of gallstone pancreatitis.

CHOLEDOCHOLITHIASIS

Obstructive jaundice due to gallstones in the common bile duct has increased.

Cholangitis is a common presenting problem with obstructive jaundice (OJ). The diagnosis is often missed and remains a clinical diagnosis with biochemical confirmation. Due to delays in presentation, patients often present with a severe form of cholangitis and have organ dysfunction, most commonly renal dysfunction.

BILE DUCT INJURIES

The increased prevalence of symptomatic gallstones implies a greater number of cholecystectomies and hence a greater number of post operative bile leaks and iatrogenic bile duct injuries. Availability of ERCP is an important issue that needs to be addressed in SSA. There is evidence to support referral to a center with the necessary skills [71].

GALLBLADDER CANCER

Gallbladder adenocarcinoma is reported as being the most common biliary tract cancer. The etiology is uncertain.

In 1995, Warren et al. reported on 19 cases over a 68-month period [72]. At CHBAH, we see this number in 3 months. They also reported a regional variation suggesting this is an important aspect to investigate further.

PARASITIC INFECTIONS OF THE BILIARY TREE

Most of these infections affect the LMICs. It is estimated that approximately 25% of the developing world's population is infested with *Ascaris lumbricoides* [73]. Biliary parasites can cause biliary obstruction, biliary lithiasis, hemobilia, sclerosing cholangitis, and cholangiocarcinoma. Ascaris may also migrate into the pancreatic duct and cause acute pancreatitis particularly in the pediatric age group.

OBSTRUCTIVE JAUNDICE (OJ)

Patients with OJ have a higher morbidity and mortality rate and as such OJ must be excluded in all patients presenting with jaundice.

The evaluation of the underlying pathology usually requires advanced investigations including sonar, cross-sectional imaging including CT or MRI, MRCP and today EUS is very helpful but its availability is limited to major centers. The further management will be based on the etiology and prompt referral to a center, which has the necessary resources and skills to complete the definitive management.

CONCLUSIONS

Biliary tract disorders are common in Sub-Saharan Africa. There are limited databases that can accurately reflect the spectrum of disease, but the extent of the pathology is well highlighted here. The late presentation of patients increases the morbidity and mortality and further expertise is currently being developed to address the shortfalls outside centers of excellence.

It is emphasized that this represents data from a center of excellence and underscores the necessity of developing centers of excellence in SSA to facilitate health care quality.

VIRAL HEPATITIS IN SSA

Viral hepatitis remains a substantial health issue in SSA and the prevalence is mostly underestimated. This burden is compounded by the lack of screening and access to care and treatment as well as inadequate disease surveillance, human and financial resources [74].

HEPATITIS A (HAV)

SSA is a hyperendemic area for HAV with most primary infections occurring before the age of 10 years. The incidence of HAV is strongly correlated with access to clean water and adequate sanitation, and this factor has mostly contributed to primary infections occurring in children. Work from West Africa has demonstrated that prior to 1990, most preschool children had evidence of immunity in keeping with the known epidemiology of most infections occurring before age 5. However, seroprevalence data from urban children have indicated that the midpoint of children population immunity has shifted to school-aged children suggesting a transition to lower levels of HAV endemicity. An optimistic feature is that 64% of people in SSA now have access to improved drinking water sources. However, significant numbers are still reliant on unimproved or surface water that is fecally contaminated. Thus, we may be seeing the first shifts in the epidemiology of water-borne diseases such as hepatitis. This epidemiological transition with changing socioeconomic circumstances as well as the high HIV burden warrants an urgent review of policy with respect to hepatitis A vaccination in immunization schedules in the near future. In SSA, given its endemicity, outbreaks of HAV are uncommon and hence routine vaccination for HAV has not been recommended. Rather, the World Health Organization has suggested that individual countries should assess their hepatitis A IgG seroprevalence status and consider inclusion of the vaccine into their own immunization schedules based on their data. This issue however should be re-evaluated in the African context given socioeconomic transitions.

HEPATITIS B (HBV)

HBV is an important public health issue in SSA with the HIV/AIDS pandemic that has ravaged the continent, complicating the natural history of HBV. Singly, it is responsible for 70% of the cases of hepatocellular carcinoma that occurs in Africa and of even greater importance, it is wholly preventable through a comprehensive strategy to manage hepatitis B including a birth-dose-based vaccination policy and a mother-to-child transmission reducing strategy. Identifying patients with chronic HBV, treating them if required and monitoring for HCC, completes the strategy.

A significant challenge is to develop cheap yet effective and good quality diagnostic tests that can be used to evaluate large populations and to determine the need for therapy. However, before this can even be achieved, hepatitis B needs to be regarded as a priority in SSA. Seroprevalence data are very limited and no policy exists where testing for HBsAg is performed as part of routine antenatal care, for example. Until screening and identification of those chronically infected is prioritized, little is likely to change.

HEPATITIS C

Worldwide approximately 130 million to 150 million people are infected with hepatitis C virus (HCV). As yet no preventive or protective vaccine exists for HCV; hence, measures to control the epidemic are focused on the known risks of HCV acquisition. Improved blood transfusion donor screening has all but eradicated this as a means of HCV transmission and needle exchange programs and opiate substitution has reduced risk from injecting drug use. This however mostly applies to HICs. The epidemiology of HCV in SSA is incompletely understood with accurate epidemiological information being very limited. Furthermore, meager data on the burden of coinfection with HIV exists. A 2002 review of HCV epidemiology in SSA estimated the overall prevalence to be 3%, although some reports place the prevalence at 5.3%. Significant regional variation has been observed. A recent meta-analysis has demonstrated this wide regional variation in seroprevalence with 6.8% in the central Africa region; 4.3% in West Africa; and 0·9% in southeast Africa.

Thus more people in Africa are infected with HCV than have HIV, yet the public health burden of HCV has only now started to be recognized. For HCV, the two major challenges for the 21st century are understanding the epidemiology in SSA so that targeted screening can occur to identify those infected as well as getting affordable access to the new direct acting antiviral (DAA) drugs that offer almost universal cure using short duration regimens.

Puzzlingly, when one broadly views a map of Africa, there is a seemingly inverse relationship between HIV and HCV prevalence with HIV prevalence being high in Southern Africa and HCV prevalence rising as one moves more north in Africa. The reasons for this are not entirely clear.

HEPATITIS D

HDV is a unique RNA virus that is dependent on HBV for survival and thus for infection. HDV is a defective virus; it does not encode its own replicase and is totally dependent on the host replicative machinery for its synthesis. Thus, there are no viral replicative enzymes for drugs to target.

HEPATITIS E

Hepatitis E, an RNA virus, was only formally identified in 1983.

Hepatitis E infects about 20 million people annually with a 0.2–4% mortality rate causing approximately 70,000 deaths. It is predominantly encountered in Southern Asia and parts of West, North-East, and Southern Africa. There is one serotype of hepatitis E and 4 genotypes, 1–4. Genotypes 1 and 2 are responsible for epidemic outbreaks and is mostly associated with contaminated water although person-to-person transmission is described. Infection is restricted to humans with genotype 1 predominating. In Africa, most outbreaks are sporadic with seroprevalence data from Zambia suggesting an overall seroprevalence of 42%; 28% in HIV-negative and 71% in HIV-positive adults. A recent study in the Western Cape in South Africa in both adults and children found an overall seroprevalence of approximately 30%.

Given its association with contaminated water, the issues around access to safe and clean water sources for preventing HAV are synonymous for hepatitis E. Given genotype 3 and its association with raw or undercooked meat, mostly pork, basic hygiene principles of avoiding undercooked meats are critical to prevention.

HEPATOCELLULAR CARCINOMA (HCC)

HCC is the sixth most common cause of cancer and the third leading cause of cancer deaths worldwide. In men, HCC is the second most common cause of death from cancer. The annual fatality ratio (0.93) is the highest of any malignancy.

SSA is one of the three geographical areas where HCC occurs very commonly. In SSA, HCC is the second most common cause of cancer in men. Viral hepatitis is the most common risk factor associated with the development of this malignancy. Screening for viral hepatitis in Africa is underutilized. Aflatoxin exposure and alcohol have a multiplicative impact on the risk of developing this cancer. Diagnosis of HCC in Africa is primarily by ultrasound scanning. Efforts should be focused on ensuring expanded access to HBV vaccination and offering HCV treatment to those who need it. Operational research is urgently needed to understand how best to screen and treat this malignancy in an African context.

In particular, improving access to prevention and treatment of viral hepatitis in Africa may herald dramatic reductions in the incidence of this malignancy over the next few decades.

EPIDEMIOLOGY IN SSA

To date, the data describing the occurrence of HCC in sub-Saharan Africa have neither been complete, nor always accurate. As a result, the incidence of HCC is likely to be significantly understated, particularly as a tissue diagnosis of HCC is rarely performed on the African subcontinent. The reports of hospital-based series of patients have been and remain biased by the local experience. Registry data are inaccurate and often not available at all. An age-standardized ratio of >15 per 100,000 of the population per year is regarded as high incidence of HCC. The age-standardized incidence rate in SSA is 41.2/100,000 persons/year: for South Africa it is 16.0 for males and 4.5 for females, for Zimbabwe 30.7 for men and 17.1 for females, and in Swaziland 22.0 in males and 5.2 in females [75]. The usual age distribution in HICs is late fifties to late seventies. In SSA, there is a distinct shift toward a younger age in Black males with a mean age of presentation around 45 years. This produces a substantial burden on health care resources as well as a drain in intellectual and productive capacity, in an area of the world that can ill afford it. Men have a higher incidence of HCC in SSA than do women in all populations. The higher incidence of HCC in this population probably reflects the higher rate of chronic HBV (approximately twice that in females).

REFERENCES

[1] Kachala R. Systematic review: epidemiology of oesophageal cancer in Sub-Saharan Africa. Malawi Med J 2010;22(3):65–70.
[2] Sitas F, Parkin M, Chirenje Z, et al. Cancers. In: Jamison DT, Makgoba MW, editors. Disease and mortality in Sub-Saharan Africa. 2nd ed. Washington, DC: World Bank; 2006.
[3] Mark T, Edwin F, Aniteye E, et al. Changing epidemiology of oesophageal cancer in Sub-Saharan Africa. Pan Afr Med J 2012;13–6.
[4] Day N, Munoz N. Esophagus. In: Schottenfield D, Fraumens J, editors. Cancer epidemiology and prevention. Philadelphia: WB Saunders; 2006. p. 596.
[5] Sharma P, Wani S, Romero I, et al. Racial and geographic issues in gastroesophageal reflux disease. Am J Gastroenterol 2008;103(11):2669–80.
[6] Segal I, Solomon A, Ou Tim L, et al. Hiatus hernia in Johannesburg Blacks. S Afr Med J 1980;58:404–5.
[7] Segal I. The gastro-oesophageal reflux disease complex in sub-Saharan Africa. Eur J Cancer Prev 2001;(3):2009–12. june 10.
[8] Segal I, Noormohamed AM, Ranchod S, et al. Duodenal and gastric ulcer in in Soweto. S Afr Med J 1983;64:777–8.
[9] Jemilohun AC, Otegbayo JA, Ola SO, et al. Prevalence of *Helicobacter pylori* among Nigerian patients with dyspepsia in Ibadan. Pan Afr Med J 2010;6.
[10] Ndububa D, Agbakwuru A, Adebayo R, et al. Upper gastrointestinal findings and incidence of *Helicobacter pylori* infection among Nigerian patients with dyspepsia. West Afr J Med 2001;20:140–5.
[11] Nwokediuko SC, Ijoma U, Obienu O, et al. Time trends of upper gastrointestinal diseases in Nigeria. Ann Gastroenterol 2012;25(1):52–6.
[12] Ameh EA, Nmadi PT. Pattern of peptic ulcer disease in Zaria Nigeria. East Afr Med J 1998;75(2):90–2.
[13] Kidd M, Louw JA, Marks IN. *Helicobacter pylori* in Africa: observations on an 'enigma within an enigma'. J Gastroenterol Hepatol 1999;14:851–8.
[14] Duggan JM, Dobson AJ, Johnson JH, et al. Peptic ulcer and non-steroidal anti-inflammatory agents. Gut 1986;27:929–33.
[15] Drini M. Peptic ulcer disease and non-steroidal anti-inflammatory agents. Aust Prescr 2017;40:91–3.
[16] Sitas F, Parkin M, Chirenje Z, et al. Cancers. In: Jamison Feachem RG, Makgoba MW, et al., editors. Disease and mortality in Sub-Saharan Africa. 2nd ed. Washington, DC: The International Bank for Reconstruction and Development/The World Bank; 2006 [chapter 20].
[17] Wabinga HR. *Helicobacter pylori* and histopathological changes of gastric mucosa in Uganda population with varying prevalence of stomach cancer. Afr Health Sci 2005;5:234–7.
[18] Segal I, Ally R, Mitchell H. *Helicobacter pylori*—an African perspective. Q J Med 2001;94:561–5.
[19] Tovey F. Duodenal ulcer in Black populations in Africa south of the Sahara. Gut 1975;16:564–76.
[20] Segal I, Noormohamed AM, Rachod S, et al. Duodenal and gastric ulcer in Soweto. S Afr Med J 1983;64:777–8.
[21] Levenstein S. Stress and peptic ulcer: life beyond helicobacter. BMJ 1998;16:538–41.
[22] John KD, Segal I, Hassan H, et al. Acute pancreatitis in Soweton Africans. A disease with high mortality and morbidity. Int J Pancreatol 1997;21(2):149–55.
[23] Ugwu BT, Obekpa PO, Kidmas AT. Acute pancreatitis at Jos university Teaching Hospital Nigeria. J Surg 1997;4(2):58–62.

[24] Segal I. Pancreatitis in Soweto, South Africa. Focus on alcohol-related disease. Digestion 1998;(suppl. 4): 25–35.
[25] Segal I, Lerios M, Mac Phail AP, et al. The genesis of chronic pancreatitis in the South African Black population. S Afr Med J 1988;74:384–6.
[26] Segal I, Lerios M, Grieve T. The emergence of chronic calcifying pancreatitis in a developing country. In: Gyr K, Singer MV, Sarles H, editors. Pancreatitis: concepts and classification. Amsterdam: Excerpta Medica; 1984. p. 417–25.
[27] Roberts SE, Akbari A, Thorne K, et al. The incidence of acute pancreatitis: impact of social deprivation, alcohol consumption, seasonal and demographic factors. Aliment Pharmacol Ther 2013;38(5):539–48.
[28] Udayakamar N, Jayanthi V. Chronic pancreatitis in India: the changing spectrum. Postgrad Med J 2007;83(983):562–3.
[29] Segal I, Paterson A, Walker ARP. Characteristics and occurrence of appendicitis in the blacks in Johannesburg, South Africa. J Clin Gastroenterol 1986;8:530–3.
[30] Saad AA, Makama JG, Mohammed U, et al. Epidemiology of appendicitis in northern Nigeria: a 10 year review. Sub-Saharan Afr J Med 2014;1(4):1185–90.
[31] Yang E, Kahn D, Cook C. Acute appendicitis in South Africa: a systematic review. S Afr J Surg 2015;53(3–4):31–8.
[32] Alatise OI, Ogunweide T. Acute appendicitis: incidence and management in Nigeria. IFEMED J 2008;14(1):66–70.
[33] Mollie F, Quan S, Kaplan B, et al. The global incidence of appendicitis: a systematic review of population-based studies. Ann Surg 2017;266(2):237–41.
[34] Segal I. Intestinal tuberculosis, Crohn's disease and ulcerative colitis in an urban Black population. SA Medizinhist J 1984;65:17–44.
[35] WGO. Inflammatory bowel disease: a global perspective. In: WGO Guidelines; June 2009. p. 2–3. Introduction. IBD.
[36] Ukwenya AA, Odige VI, et al. Inflammatory bowel disease in Nigerians. Still a rare diagnosis? Ann Afr Med 2011;10(2):175–9.
[37] Epstein D, Watermeyer G, Kirsch R. Review article: the diagnosis and management of Crohn's disease in populations with high-risk rates for tuberculosis. Aliment Pharmacol Ther 2007;25:1373–8.
[38] Epstein D. Inflammatory bowel disease in Cape Town. "Experience and evidence" abstract. SAGES Congress; 2014.
[39] Chu KM, Watermeyer G, Shelly L, et al. Childhood helminth disorder is protective against inflammatory bowel disease: a case control study in South Africa. Inflamm Bowel Dis 2013;19:614–20.
[40] Basson A, Swart R, Jordaan E, et al. The association between race and Crohn's disease phenotype in the Western Cape population of South Africa, defined by the Montreal classification system. PLoS One 2014;9(8):e104859.
[41] Deetlefs E, Epstein D, Watermeyer G, et al. Tuberculosis in an inflammatory bowel disease cohort from South Africa. S Afr Med J 2012;102(10):802–4.
[42] Matrana MR, Margolin DA. Epidemiology and pathophysiology of diverticular disease. Clin Colon Rectal Surg 2009 Aug;22(3):141–6.
[43] World Gastroenterology Organisation Practice Guidelines. Diverticular Disease, http://worldgastroenterology.org//global-guidelines-guidelines/diverticular-disease/diverticular-disease-english.
[44] Segal I, Solomon A, Hunt J. Emergence of diverticular disease in the urban South African Black. Gastroenterology 1977;72:215–9.
[45] Vally M, Koto MZ, Govender M. An investigation of diverticular disease among black patients at George Mukhari Academic Hospital, Pretoria, South Africa. S Afr Med J 2017;107(2):137–9.
[46] Oluyemi A, Odeghe E. Diverticular disease at colonoscopy in Lagos State, Nigeria. Niger Med J 2016;57(2):110–3.

[47] Kiguli-Malwadde E, Kaszi H. Diverticular disease of the colon in Kampala, Uganda. Afr Med Sci 2002;2(1):29–32.
[48] Ihekwaba FN. Diverticular disease of the colon in Black Africa. J R Coll Surg Edinb 1992;3:107–9.
[49] Mehari H. Management of sigmoid volvulus in Eritrea, www.unige.ch/cyberdocuments/theses2002/MehariH/these_body.html.
[50] Gingold D, Murrell Z. Management of colonic volvulus. Clin Colon Rectal Surg 2012;25(4):236–44.
[51] Segal I. Decreasing incidence of sigmoid volvulus at Baragwanath Hospital. S Afr Med J 1985;67:443.
[52] Jemal A, Bray F, Center M, Ferlay J, Ward E, Forman D. Global cancer statistics. CA Cancer J Clin 2011;61(2):69–90.
[53] Cappell MS. Pathophysiology, clinical presentation, and management of colon cancer. Gastroenterol Clin N Am 2008;37(1):1–24.
[54] Madiba T. Colorectal cancer. In: Adeloye A, Adekunle O, Awojobi O, editors. Davey's Companion To Surgery In Africa. Eruwa, Nigeria: Acecool; 2009.
[55] Parkin D, Ferlay J, Hamdi-Chérif M, et al. Cancer in Africa. Lyon: IARC Scientific Publication; 2003.
[56] Graham A, Davies Adeloye LG, Theodoratou E, Campbell H. Estimating the incidence of colorectal cancer in Sub-Saharan Africa: a systematic analysis. J Global Health 2012;2(2):1–14.
[57] Kauh J, Brawley O, Berger M. Racial disparities in colorectal cancer. Curr Probl Cancer 2007;31:123–33.
[58] Béjar LM, Gili M, Infantes B, Marcott PF. Effects of changes in dietary habits on colorectal cancer incidence in twenty countries from four continents during the period 1971–2002. Rev Esp Enferm Dig 2011;103(10):519.
[59] Saidi H, Nyaim EO, Githaiga JW, Karuri D. CRC surgery trends in Kenya, 1993–2005. World J Surg 2008;32(2):217–23.
[60] Segal I, Edwards CA, Walker AR. Continuing low colon cancer incidence in African populations. Am J Gastroenterol 2000;95(4):859–60.
[61] Munyaradzi OM. Carcinoma of the stomach and large bowel—a comparative study of the incidence in blacks and whites in Zimbabwe/Rhodesia. S Afr J Surg 1979;17(4):179–87.
[62] Angelo N, Dreyer L. Colorectal carcinoma--a new threat to black patients? A retrospective analysis of colorectal carcinoma received by the Institute for Pathology, University of Pretoria. S Afr Med J 2001;91(8):689–93.
[63] Bonnet GC, Grobler SP. Disease profile of colon and rectum carcinoma at Pelonomi Hospital Bloemfontein. S Afr J Surg 1997;35(2):63–5 [discussion 5–7].
[64] Boytchev H, Marcovic S, Oettle GJ. The characteristics of large bowel cancer in the low-risk black population of the Witwatersrand. J R Coll Surg Edinb 1999;44(6):366–70.
[65] Degiannis E, Sliwa K, Levy R, Hale MJ, Saadia R. Clinicopathological trends in colorectal carcinoma in a Black South African population. Trop Gastroenterol 1995;16(4):55–61.
[66] Madiba T, Sartorius B, Ramdial P. Clinicopathologic spectrum of colorectal carcinoma in the KwaZulu-Natal Province of South Africa: lessons from a colorectal cancer database. S Afr J Sci 2015;111.
[67] Beyers CF. Incidence of surgical diseases among the Bantu Races of South Africa. S Afr Med J 1927;1:606–12.
[68] Brebner IW. Surgical conditions the liver including the Gallbaldder. S Afr Med J 1934;8:867–71.
[69] Walker AR, Segal I, et al. Prevalence of gallstones in elderly black woman in Soweto, Johannesburg, as assessed by ultrasound. Am J Gastroenterol 1989;84:1383–5.
[70] Parekh D, Lawson HH, Kuyl JM. Gallstone disease among black South Africans. A review of the Baragwanath Hospital experience. S Afr Med J 1987;72(1):23–6. Gastroenterol 1989;84:1383–5.
[71] Mercado MA, Franssen B, Dominguez I, et al. Transition from a low: to a high-volume centre for bile duct repair: changes in technique and improved outcome. HPB (Oxford) 2011;13(11):767–73.
[72] Warren BL, Marais AW. Carcinoma of the Gallbladder—a possible regional predisposition in the western Cape and northern Cape. S Afr J Surg 1995;33(4):161–4.
[73] Yellin AE, Donovan AJ. Biliary lithiasis and helminthiasis. Am J Surg 1981;142:128–36.

[74] Note this section has been extracted from a chapter in this book entitled: Sonderup M., Spearman W. Aspects of viral hepatitis in sub-Saharan Africa in the 21st century.
[75] This is extracted from a chapter in this book entitled: Setshedi M, Davisson M. Hepatocellular carcinoma in Africa.

FURTHER READING

[76] Balakrishnan V, Nair P, Radhakrishnan L, Narayanan VP. Tropical pancreatitis-a distinct entity, or merely a type of chronic pancreatitis? Indian J Gastroenterol 2006;25(2):74–81.

CHAPTER

EMERGING NONINTESTINAL DISEASES ARISING IN SSA

5

Isidor Segal

G.I.T. Unit, Chris Hani Baragwanath Hospital and University of Witwatersrand, Soweto, South Africa

OBESITY

Economic advancement has been associated with an increase in obesity and Type 2 diabetes in SSA. In 1990, Walker et al. in a study of obesity in Black women showed that in this low socioeconomic and low-fat dietary context, obesity in the Black women studied was not specifically evocative of deleterious sequelae of obesity. Minimal effects on hypertension, hyperlipidemia, and hyperglycemia were observed [1]. Later Agyemang et al. [2] noted that overweight and obesity rates are increasing in all African regions, with the Southern African region the most affected. Vorster [3] showed that with the nutritional transition, inter alia, there has been an increase in overnutrition, overweight, and obesity. The rate of overweight and obesity is higher among women than men and in and in urban areas compared to rural areas. Socioeconomic status, age, parity, marital status, physical inactivity, body weight perceptions, and increased energy intake are strong predictors of overweight and obesity in SSA. This is likely to exacerbate cardiovascular disease and diabetes.

CARDIOVASCULAR DISEASE (CVD) [4]

An epidemic of CVD is occurring in SSA. The spectrum covers stroke, heart failure, and kidney disease. At least two-thirds of CVD deaths occur in LMICs. Hypertension is the commonest risk factor.

These diseases have been present for many years but no preventive strategies, lack of resource and health care systems, poor facilities, often minimal patient follow up, and lack of sustainable drug therapies often lead to a feeling of helplessness among Health Care Workers.

It is noted that high blood pressure, tobacco smoking, history of hypertension or diabetes, nutritional factors such as high salt and low fruit intake, obesity, excessive alcohol consumption, lack of exercise, raised blood lipids, and psychosocial factors are major risk factors.

The morbidity and mortality among people, often in their most productive years, is devastating and adds to the economic impasse in these countries.

Digestive Diseases in Sub-Saharan Africa. https://doi.org/10.1016/B978-0-12-815677-3.00005-0

DIABETES MELLITUS [5]

Diabetes in the LMICs of SSA is a devastating disease that affects younger people than in HICs and has substantial economic effects. As with many other diseases in SSA, there is a lack of reliable evidence that reflects the enormity of diabetes and its complications on the population. Also the interventions in SSA will be apposite in order to alleviate the problem.

The prevalence in 2014 was concerning (7.1%—up 129% since 1980 WHO Report). Thus, countries in SSA are facing a major public health crisis in a continent with competing priorities and limited health resources.

Unfortunately, diabetes care is generally poor—there is a lack of trained health care workers and a paucity of services such as transport for poor people to obtain access to hospital services. Dietary recommendations are difficult to adhere to as the basic staples are carbohydrate and limited incomes make it difficult to follow dietary guidelines.

Other major problems are that only half of people with diabetes are diagnosed and only a small proportion of patients (11%; range: 7–33) are able to persevere with long-term follow up.

MENTAL DISORDERS

Mental disorders are among the ten top causes of disability in SSA. The range of diseases include acute toxic psychosis, often related to home brews with additives such as western spirits and other toxins, locally distilled liquor, cannabis and other illicit drugs.

Paranoid psychosis; depression, psychosis related to HIV; and cerebral involvement due to other infectious diseases such as malaria and typhoid fever are common among SSA populations.

Epilepsy has a high prevalence and is highly stigmatized in many African countries [6].

Half the population of SSA comprises children below 15 years of age. It is estimated that of those aged 0–9 years approximately 3% have a mental disorder [6].

Other causes of mental disorders are internecine strife, a flood of refugees escaping wars and natural disasters such as drought, famine, and floods.

Overall in SSA, mental health care is poor and facilities limited. The European Region of the WHO has more than 86,000 psychiatrists and 280,000 psychiatric nurses serving a population of 840 million, while SSA has about 1200 psychiatrists and 12,000 psychiatric nurses for a population of about 620 million [6].

Public awareness campaigns are hampered by the belief of many Africans in traditional healers. A study by Segal [7] states that the traditional healer wields the greatest and most awesome power and that most people consult traditional healers before doctors. It was estimated that approximately 80% of Sowetans (an urbanized Black population in Johannesburg) consult these healers. They act as psychologists, psychiatrists, marriage counselors, and healers [7].

This illustrates the difficulties in tackling the problem.

WHO and the World Psychiatric Association suggest that action is necessary to combat the situation. Programs are necessary to have a reliable information system, planning of the allocation of resources, training of psychiatrists and psychiatric nurses and health workers.

Creation of national psychiatric associations, organization of scientific meetings within regional and international scientific activities to bring forward the mental health concerns of the continent [6].

EXTRAINTESTINAL CANCERS IN SSA

Cancer is an emerging health problem in the low- and middle-income countries of SSA. In 2012, 847,000 new cancer cases and 591,000 deaths were reported [8,9]. Moreover, other cancers such as lung, breast, and prostate are diagnosed more frequently than in the past [9]. The reasons include aging, population growth, and improved detection services. The major reasons for the increase and change in profile of cancer in SSA are urbanization with its accompaniments of xenobiotics, air pollution, occupational hazards, increased alcohol consumption, cigarette smoking, and dietary changes. A high prevalence of infectious agents, e.g., HIV/AIDS, human papilloma virus are also responsible for the increase [10].

While the cancer profiles often differ widely between regions, the most common cancers in females were breast and cervical cancer (50%). In males cancer of the prostate predominates (27.9%; 51,900 cases). Other less common cancers in men are Liver (10.4%) and Kaposi sarcoma (6.7%) [10].

The numbers of cancers are expected to double in the next 20 years.

Unfortunately, in most African countries, cancer control programs, cancer registries, provision of specialized detection, and treatment services are poor. This includes basic services such as vaccinations for liver and cervical cancer [8]. Public health policies to discourage smoking and alcoholism are not off the ground.

REFERENCES

[1] Walker ARP, Walker BF, Manetsi B, et al. Obesity in black women in Soweto, South Africa: minimal effects on hypertension, hyperlipidaemia and hyperglycaemia. J R Soc Health 1990 Jun;110(3):101–3.
[2] Agyemang C, Boetemaa S, Frempong GA, et al. Obesity in Sub-Saharan Africa. In: Metabolic syndrome. Switzerland: Springer International Publishing; 2016. p. 41–53.
[3] Vorster HH. Changes in nutritional status of Africans and dietary recommendations for Sub-Saharan populations. (Chapter 6. Digestive Diseases in Sub-Saharan Africa. Changes and Challenges).
[4] Cappuchio FP, Miller MA. Cardiovascular disease in Sub-Saharan Africa: burden, risk and intervention. Intern Emerg Med 2016;11:299–305.
[5] Jaffar S, Gill G. The crisis of diabetes in sub-Saharan Africa. Lancet Diabetes Endocrinol 2017;5(8):574–5.
[6] Okasha A. Mental health in Africa: the role of the WPA. World Psychiatry 2002;1(1):32–5.
[7] Segal I, Ou Tim L. The witchdoctor and the bowel. S Afr Med J 1979;36:308–10.
[8] Jemal A, Bray F, Forman D, et al. Cancer burden in Africa and opportunities for prevention. Cancer 2012;118(18):4372–84.
[9] Parkin DM, Bray FJ, et al. Cancer in Africa 2012. Cancer Epidemiol Biomark Prev 2014;23(6):953–66.
[10] Cancer Atlas n.d. Cancer in Sub-Saharan Africa. The Cancer Atlas; http://canceratlas.cancer.org/the-burden-of-cancer-in-subsaharan-africa, Am Ca Soc: IARC, WHO, UICC.

CHAPTER

6

CHANGES IN NUTRITIONAL STATUS OF AFRICANS: DIETARY RECOMMENDATIONS FOR SUB-SAHARAN AFRICA

Hester H. Vorster

Head of Centre of Excellence for Nutrition, Faculty of Health Sciences, North-West University, Potchefstroom, South Africa

INTRODUCTION

The gastrointestinal (GI) system is optimally designed and intricately regulated by neurological and hormonal systems to take in food and beverages, to extract nutrients and other substances from food, to transport nutrients and these substances through complex mechanisms to the circulation for further delivery and use by all tissues, and finally to excrete waste products from the body. It is, therefore, not surprising that what we eat and drink, and consequent nutritional status, have a huge impact on the physiology and pathology of the GI system. Moreover, the health of the GI system influences how well humans can respond to what they eat and how they will cope with dietary changes. The relationship of diet and the GI system is therefore not only a therapeutic one, but also preventative: balanced, adequate, or optimal nutrition is necessary to maintain a healthy gut.

We live in an age of change, to such an extent that we can probably define health as "our ability to cope with change." The gradual, transitional changes in diet and lifestyle that accompanied "ecological" development of man in the "western" world, and consequent disease patterns or epidemiology, is happening at an alarming accelerated rate in sub-Saharan Africa (sSA). This is a result of the end of colonialism in Africa, political freedom, modern communication technologies, and the sudden availability, access to and affordability of foods, ingredients and drinks not previously consumed by large sections of African populations.

These changes in food and nutrient intakes are known as the nutrition transition (NT). The NT, together with other behavioral changes, are responsible for epidemiological changes, characterized by decreased fertility, and decreases in infectious- and nutrition deficiency diseases, accompanied by an increase in noncommunicable diseases (NCDs) observed in developing countries all over the world.

For example, in their 2013–2020 Action Plan for prevention of global NCDs, the World Health Organization (WHO) reported that globally, NCDs cause >14 million premature deaths between the ages of 30 and 70 each year. Of concern is that 85% of these deaths occur in developing countries, including sSA.

Digestive Diseases in Sub-Saharan Africa. https://doi.org/10.1016/B978-0-12-815677-3.00006-2

Poverty and a lack of infrastructure and human resources to treat NCDs in developing countries is of course a major reason for the high mortality. But another reason is that the NT has hit the poorer developing countries especially hard, because it happened before the battle against undernutrition has been won, and also because it happened at an accelerated rate. Whereas the dietary changes characteristic of the NT in Europe and North America took centuries to reach its present state, it happened almost overnight, in a few decades, in sSA. This unfortunate situation is further exacerbated by the HIV/AIDS epidemic, and the periodic outbreaks of Ebola and other infectious xenobiotic diseases.

To understand and learn from the pathophysiology of GI diseases in sSA, it would be worthwhile to have a closer look at the NT and coexistence of under- and overnutrition, its causes, consequences, and how this transition should influence dietary recommendations for prevention of disease in the region, as well as other regions experiencing such a rapid NT.

THE NUTRITION TRANSITION (NT): CHARACTERISTICS, CONSEQUENCES, AND CAUSES

CHARACTERISTICS OF DIETARY AND NUTRIENT CHANGES

In sSA, "political freedom," urbanization, modernization (in urban and now also in rural areas) as well as acculturation, all lead to changing lifestyles. These include emancipation of women (more women in the work force), more "previously disadvantaged" people in managerial and high political positions and offices and therefore increasing stress levels, increase of the smoking habit, less activity, and much more palatable and acceptable diets.

The dietary changes are basically because of the increased availability (and often aggressive marketing) of cheaper meat (chicken), plant oils, sugar, ultraprocessed "junk" products (I hesitate to call them foods), snacks, and sugar-sweetened beverages. The changes in diet and resultant nutrient intakes have been documented in some detail in South Africa and also observed in other African countries. Despite some country-specific and cultural differences, they are essentially similar to those described for other regions such as South America, India, and China. These changes are shown in Table 6.1.

It was the high dietary fiber intake of rural Africans and the low prevalence of NCDs that originally lead to the hypothesis that the diet of rural Africans protects them against many chronic noncommunicable diseases. The changes in dietary fiber intake with urbanization is intriguing: If people move from arid regions to urban cities, they may actually increase dietary fiber intake if they can afford to eat more vegetables and fruit, as we have shown in the North-West province of South Africa. In this region, rural dwellers often has cattle, sheep, and goats, but little water to irrigate vegetable gardens. However, a consistent decrease of staple foods and thus starch and especially resistant starch that is fermentable in the colon, could lead to an increase in gastrointestinal diseases.

CONSEQUENCES

A major consequence of the rapid NT in sSA is the coexistence of under- and overnutrition, leading to the typical double burden of disease.

The impact of dietary changes showed in Table 6.1 on nutrient intakes and some health outcomes of Africans are illustrated in Table 6.2. The data come from one of the earlier studies, which examined the NT from 1996 to 1998, namely, the THUSA study in the North-West Province of South Africa.

Table 6.1 Dietary and Nutrient Intake Change Associated With the Nutrition Transition (NT)

Change in Food Intake: Modernized Dietary Patterns	Change in Nutrient Intakes (Energy-Dense and Micronutrient-Poor Diet) and Some Outcomes	Underlying Reason(s)
More chicken, red meat, eggs	More animal protein and saturated fat; more iron, zinc, and vitamin B_{12}. Improvement of undernutrition	Increased availability and cheaper chicken
More fried, salty foods	More omega-6 fatty acids (FA), salt, and total fat, but less omega-3 FA. Increased risk of NCDs	Cheaper plant oils; decreases in traditionally local foods
Less wholegrain starchy staples	Less total carbohydrate (CHO); less dietary fiber; less micronutrients. Constipation and gut diseases increase	Fashionable to have white, purified grains
More sugar (sucrose): sweets, sweetened beverages, cakes, puddings, cookies, etc.	More refined CHO (sucrose), despite less total CHO. Increased risk of NCDs	Cheap sugar available, and affordable; acceptability of sweet taste
More ultraprocessed packaged foods and drinks;	More sugar, salt, fat, energy, less micronutrients. Overweight, obesity, and NCDs increase	Working housewives (househusbands); less time for food preparation
Less vegetables: less vegetable gardens and fresh home-prepared vegetables	Less dietary fiber, less micronutrients: hidden hunger and increased risk of NCDs	Convenience, less time to cook, less space for vegetable gardens
Larger serving sizes	More energy (from all macronutrients): overweight, obesity, and increased NCDs	Sociological and psychological reasons; affordability of food
Less breast feeding	Overconsumption of all nutrients by infants; unfavorable epigenetic profile, rapid growth rates, and obese children	Convenience and affordability of infant formula foods

In this study, the effect of urbanization on the health of Africans was examined by comparing nutrition and health variables across 5 levels of urbanization in "apparently" healthy men and women.

The selected data in Table 6.2 show how the changes in dietary energy distribution are accompanied by increases in some risk factors of NCDs. Note that in women, mean body mass index (BMI) rapidly changed from normal (BMI = 20–24.9 kg/m^2) to overweight (25–29.9 kg/m^2 with urbanization.

The double burden (coexistence of under- and overnutrition) is illustrated in Table 6.3, showing the percentage of deaths and disability adjusted life years (DALYS) extracted from an estimation by the South African Medical Research Council from data of the year 2000. Of the 17 risk factors, 5 are related to overnutrition, and 4 to undernutrition. This double burden phenomenon is also illustrated by the prevalence of undernutrition (stunting) among children and overnutrition (overweight and obesity) among adults in South Africa. For example, the recent (2012/13) South African National Health and Nutrition Examination Survey (SANHANES) showed that 26.5% of 1–3 year old children were stunted, suffering from chronic undernutrition, while 55% of adult women aged 45–54, were obese (BMI > 30 kg/m^2) and another 20% overweight (BMI: 25–29.9 kg/m^2). Thus, 75% of women were overnourished while a quarter of SA children are still chronically undernourished.

Table 6.2 Mean (Standard Deviation or 95% Confidence Intervals) of Dietary Intakes and Health Variables of Africans Across Different Levels of Urbanization[a]

Variable		Level 1[a]	Level 2[a]	Level 3[a]	Level 4[a]	Level 5[a]
N women	(w)	300	148	175	293	106
N men	(m)	196	113	134	236	64
% Fat energy (*E*)	(w)	23.6 (7.1)	22.5 (6.7)	25.7 (7.7)	27.7 (6.8)	31.8 (5.7)
	(m)	22.9 (6.8)	22.5 (7.4)	24.2 (7.8)	26.0 (7.1)	30.6 (6.7)
% Carbohydrate *E*	(w)	67.0 (9.4)	68.4 (9.0)	64.1 (9.7)	61.5 (9.3)	55.6 (7.5)
	(m)	67.4 (9.2)	67.5 (9.7)	65.6 (10.0)	64.0 (9.4)	57.3 (8.6)
% Protein *E*	(w)	11.4 (2.0)	11.3 (2.1)	11.8 (2.0)	12.1 (2.3)	13.4 (2.3)
	(m)	11.6 (2.2)	12.1 (2.1)	11.9 (1.9)	11.8 (2.1)	13.2 (1.8)
Dietary fiber (g)	(w)	15.8 (7.3)	15.4 (6.8)	16.1 (7.8)	17.1 (7.9)	17.7 (8.1)
	(m)	19.2 (10.9)	15.6 (8.1)	17.3 (8.1)	18.8 (9.2)	19.7 (9.1)
Body mass index (kg/m^2)	(w)	25.6 (24.8–26.3)	26.3 (25.2–27.4)	26.7 (25.7–27.7)	28.0 (27.3–28.8)	28.1 (26.7–29.4)
	(m)	20.7 (20.2–21.3)	20.6 (19.9 21.3)	20.3 (19.7–20.9)	21.3 (20.8–21.8)	23.1 (22.2–24.0)
Serum cholesterol (mmol/L)	(w)	3.99 (3.88–4.10)	4.11 (3.95–4.27)	4.10 (3.95–4.25)	4.43 (4.31–4.54)	4.76 (4.57–4.96)
	(m)	3.88 (3.75–4.01)	4.10 (3.82–4.17)	3.86 (3.70–4.02)	3.98 (3.86–4.10)	4.69 (4.45–4.92)
Systolic blood pressure (mmHg)	(w)	128 (126–130)	138 (133–143)	136 (132–140)	135 (133–138)	121 (115–127)
	(m)	125 (123–127)	125 (120–129)	134 (130–137)	132 (130–135)	125 (120–130)
Diastolic blood pressure (mm/Hg)	(w)	78 (76–79)	82 (78–85)	84 (81–87)	83 (82–85)	75 (71–79)
	(m)	76 (74.4–77.4)	75.6 (72.5–78.6)	80.7 (78.1–83.3)	78.9 (77.3–80.5)	79.9 (76.1–83.6)

[a]Level 1: deep rural; Level 2: farm dwellers; Level 3: informal housing areas; Level 4: poor urban dwellers; Level 5: richer urban dwellers. Note: The lower blood pressures in group 5 may probably be because of availability of better medical care in urban areas.

POSSIBLE CAUSES OF THE DOUBLE BURDEN

The question is why do populations move from undernutrition to overnutrition and not to optimal, balanced, or healthy nutrition during the NT? The answer probably lies in several interacting factors, all important for addressing the negative consequences of the NT in sSA. These factors include a biological link between under- and overnutrition, leading to a vicious intergenerational cycle, embedded in poverty. Another group of factors are probably related to sociocultural responses to economic development.

The biological link

The vicious under- and overnutrition cycle is illustrated in Fig. 6.1, indicating that undernutrition during fetal life (because of undernourished mothers and insufficient nourishment of the fetus) could result

Table 6.3 Contribution of Selected Potentially Modifiable Risk Factors to Percentage of Deaths and DALYS in South Africa in 2000 (521 Thousand Deaths and 16.2 Million DALYS)

Identified Risk Factor	% Total Deaths	% Total DALYS
Unsafe sex/STDs (HIV/AIDS)	26.3	31.5
High blood pressure (overnutrition related)	9.0	2.4
Tobacco smoking	8.5	4.0
Alcohol harm (overconsumption related)	7.1	7.0
High body mass index (overnutrition related)	7.0	2.9
Interpersonal violence (risk factor)	6.7	8.4
High cholesterol (overnutrition related)	4.6	1.4
Diabetes (risk factor) (overnutrition related)	4.3	1.6
Physical inactivity	3.3	1.1
Low fruit and vegetable intake (undernutrition related)	3.2	1.1
Unsafe water, sanitation, and hygiene	2.6	2.6
Child and maternal underweight (undernutrition related)	2.3	2.7
Urban air pollution	0.9	0.3
Vitamin A deficiency (undernutrition related)	0.6	0.7
Indoor air pollution	0.5	0.4
Iron deficiency anemia (undernutrition related)	0.4	1.1
Lead exposure	0.3	0.4

Estimations of the SA MRC, taking into account factors such as "likely to be among the leading causes of burden of disease or injury; evidence of causality; being potentially modifiable; and availability of data."

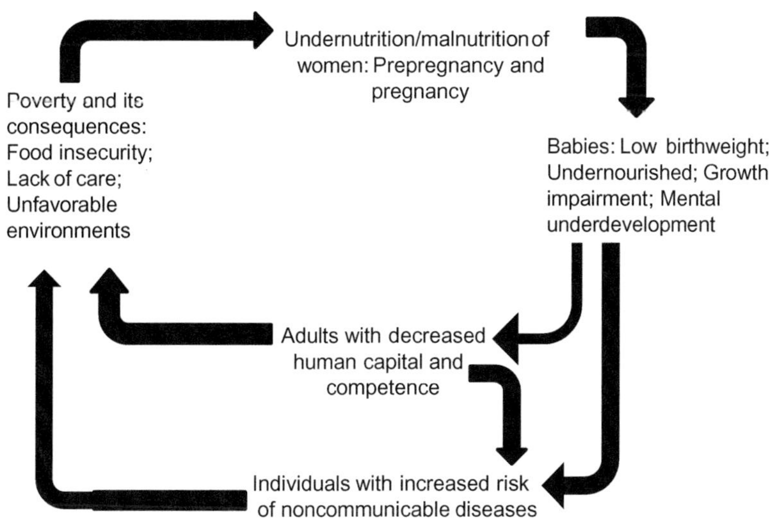

FIG. 6.1

The vicious intergenerational cycle of poverty and undernutrition.

in low-birth-weight (LBW) or disproportionate babies (with abnormal body proportions). Moreover, it could also result in babies with a changed epigenome. It has been shown that undernutrition, and possibly also other insults like the mother smoking during fetal life, could influence the epigenome of the developing fetus. Possible mechanisms include methylation of DNA, and modifications of histones and chromatin. The changed epigenome leads to changed genetic expressions in the new individual throughout its lifetime. This means that specific metabolic pathways may be activated or inhibited in different ways, increasing risks of several NCDs.

Fig. 6.1 also illustrates that undernourished mothers have babies that may be compromised regarding normal physical and mental development, and will, especially with further nutritional insults, be vulnerable to infectious diseases. They may develop into stunted infants and children who cannot fully benefit from education, and thus will have difficulty in reaching their full genetic potential. They are, at the same time, metabolically programmed (because of the changed epigenome) to be more vulnerable to NCDs in later life. Therefore, when these children are exposed to an unfavorable, "Westernized" or modern energy-dense, micronutrient-poor food environment, they are at greater risk to become overweight and obese and to develop NCDs in adult life. The combination of underdevelopment, infectious diseases, and NCDs could lead to adults with low human capital, unable to economically care for themselves and their offspring, and eventually to poverty. Poverty is one of the root causes of undernutrition, and expectant mothers from poor environments are at greater risk to give birth to babies who were nutritionally compromised in the womb, perpetuating the poverty-malnutrition cycle.

There is sound scientific evidence supporting the cycle. Pivotal in the cycle is the relationship between low human capital and poverty, the latter being defined here not only in economic terms, but also as deprivation of the capability to lead a minimally acceptable life. "Human capital" is usually seen as well-nourished, healthy, educated, skilled, and alert individuals, resulting in a labor force that could be any country's most productive asset. Reduced mental capacity and lower human capital development are results of fetal but also early infant and childhood malnutrition because of malnourishment of pregnant women, as well as further nutritional and poverty-related health insults during child development. The reduced human capital is associated with an inability to grow optimally, develop and benefit from education, poor socialization and generally a reduced capacity to lead productive lives that will ensure food and nutrition security for self and family.

It is important to note that not only undernutrition, but also maternal prepregnancy and pregnancy overweight and obesity influence birth outcomes while overnutrition during pregnancy is now known also to program the fetus for later obesity. Prepregnancy overweight increases the risk for hypertension during pregnancy, pre-eclampsia, gestational diabetes, necessity for Cesarean section, large-for-gestational age and macrosomic infants, and possibly also of birth defects such as neural tube and congenital heart disease. Less is known about the effect of prepregnancy underweight on birth outcomes, but the available data showed that balanced energy-protein supplementation during pregnancy had significant benefit. These findings emphasize the importance of avoiding both under- and overnutrition before and during pregnancy.

The sociocultural responses to economic development

Less is known about possible sociocultural reasons why so many individuals in developing populations overconsume energy when they are exposed to sufficient food. It could be that people who were previously food insecure tend to overcompensate and eat too much when food becomes available. It could also be that people are used to large portion sizes or volumes of low-energy staple foods in traditional diets, and fail to reduce portion sizes when confronted with energy-dense foods. With economic development, inexpensive sweet and oily refined and highly processed products (foods and snacks), which are energy dense but micronutrient poor (and low in dietary fiber), become available and affordable. It has been shown that man has an inherent preference for rich, creamy (oily), excessively sweet and/or salty foods. This means that developing populations have to be "educated" to maintain previously healthy eating patterns, and to make healthier choices from a changed food environment.

There are some indications that obesity in the African culture is more acceptable than in other cultures with perceptions that obese individuals are happier, more prosperous, and free from HIV/AIDS. To what extent this is true today and may contribute the obesity prevalence in African women (which is only slightly higher than in other population groups) remains to be investigated. But it is accepted that to address the problem of obesity, cultural aspects regarding both food consumption and physical activity should be considered.

It is recognized that a lack of physical activity with urbanization and acculturation, and thus energy expenditure, is part of the problem. The THUSA study (mentioned earlier) showed that lack of physical activity in African women was the greatest risk factor for obesity in these women. If increased physical activity such as walking to work, climbing stairs, and play in the case of children is not possible (perhaps because of space or safety issues), exercise and participation in sport should be promoted.

DIETARY AND POLICY RECOMMENDATIONS FOR SSA

The coexistence of under- and overnutrition in the same household, family, or community, leading to a double burden of infectious disease and high risk of NCDs, poses a major problem that should be addressed in nutrition strategies of all sSA countries.

DIETARY RECOMMENDATIONS: DEVELOPMENT OF FOOD-BASED DIETARY GUIDELINES

The nutrient needs of different gender, age, and "physiological" (pregnancy, lactation, activity level) groups differ. However, across populations, these requirements are similar. A specific nutrient profile or combination can be met by many different food combinations (or diets). Therefore, worldwide, dietary recommendations for specific regions or countries should be made using locally produced, indigenous, cultural, affordable, and available foods. People eat foods, and not nutrients. Therefore, recommendations must be made in terms of foods, as long as these foods will meet energy and nutrient requirements for a balanced diet. At the same time, the recommendations must lead to prudent intakes that will protect against overweight and high risk of NCDs. These recommendations are often in the form of a core set of food-based dietary guidelines (FBDGs), promoted by the WHO and FAO to be used by all professionals involved in nutrition education and writing of nutrition-related health policies. Needless to say, all dietary recommendations should be based on the best available evidence that they will do no harm and will meet nutrient requirements.

The key messages of the revised, qualitative FBDGs of South Africa are shown in Box 6.1. In the published documents, these guidelines (key messages) is supported by a scientific paper in which the guideline is motivated and explained, also including ways to implement it and advice on how to

BOX 6.1 REVISED GENERAL FBDGS FOR SOUTH AFRICANS (2012)

1. Enjoy a variety of foods
2. Be active!
3. Make starchy foods part of most meals
4. Eat dry beans, split-peas, lentils, and soya regularly
5. Eat plenty of vegetables and fruit every day
6. Consume low-fat milk, maas (fermented milk), or yoghurt every day
7. Lean chicken or lean meat or fish or eggs can be eaten daily
8. Drink lots of clean, safe water and make it your beverage of choice
9. Eat and use the right types of fats and oils in moderation
10. Use sugar and foods and drinks high in sugar sparingly
11. Use salt and foods high in salt sparingly

overcome barriers of implementation. The total set of guidelines also includes specific guidelines for infants and children. The guidelines are furthermore accompanied by a food guide showing the groups of foods from which daily choices should be made.

POLICY RECOMMENDATIONS

On a policy level, there should be strategies in place to:

- Improve the food environment
- Create opportunities for increased physical activity
- Inform, empower, and educate consumers to make the right choices

Improve the food environment

Modernization, economic development, and increased buying power may result in food choices from an unfortunate "westernized" food environment. As mentioned, this environment offers convenient, high-energy dense foods and snacks (rich in fats, oils, salt, and sugars), which are often low in dietary fiber and micronutrients, at increasingly affordable prices. In addition to taste preference, the aggressive marketing of these energy-dense, micronutrient-poor fast foods and peer pressure to consume them, especially when knowledge of healthful diets is lacking, probably contribute to the growing problem of obesity in developing countries.

To create an improved food environment, all stakeholders or role players should work together, with a common ethical agenda, to provide healthier food choices to consumers at affordable prices. It may even be necessary to legislate for improvement of the environment. Examples are mandatory lowering of salt and *trans*-fats in foods, prohibition of advertisements regarding certain products aimed at children during prime-time television watching, mandatory labeling of foodstuffs, as well as increased taxes on certain products, and exemption of taxes on others.

Create opportunities for increased physical activity

In addition to national programs to increase physical activity throughout the life cycle, Governments should create safe and accessible places where individuals and families can participate in exercise and sports without costs. Many developed countries already have such spaces, and it is time that sSA Governments invest in the health of their citizens by providing these opportunities to all.

Educate consumers to make the right choices

A lack of education and lack of knowledge about healthful diets of especially women and caregivers that are responsible for feeding children and procure food for the family, may play a role in both under- and overnutrition. Therefore, any national program to improve health should include components on the empowerment and education of women. Governments should promote healthy eating patterns among the total population. This can be done by using a set of FBDGs that address the nutrition-related health problems in a country, implemented in programs that use social marketing and education principles to target different groups in the population.

Educating, informing, and empowering consumers to make healthy choices from the food environment create a **demand** for a healthier food environment. It also places some of the responsibility of nutritional health on the consumers themselves. However, it is well known how difficult it is to change nutritional behavior and it is important to use a science-based approach to get balanced nutrition messages across by translating complex scientific information into meaningful health-promoting strategies that can be applied across the life course.

In conclusion, there are still some countries in sSA that can steer the NT into a more positive direction, thus avoiding the double burden of disease. However, it will need political will and allocation of sufficient resources to help the population to make the right choices to prevent NCDs, including the prevention of some gastrointestinal diseases.

RECOMMENDED READING

[1] Norman R, Bradshaw D, Schneider M, et althe South African comparative risk assessment collaborating group. A comparative risk assessment for South Africa in 2000: towards promoting health and preventing disease. S Afr Med J 2007;97(8):637–41.

[2] Sen A. The possibility of social choice, Nobel lecture. Econ Sci 1998;190:178–225.

[3] Shisana O, Labadarios D, Rehle T, et al. South African National Health and Nutrition Examination Survey. Cape Town: HSRC Press; 2013.

[4] Solomons NW. Developmental origins of health and disease: concepts, caveats, and consequences for public health nutrition. Nutr Rev 2009;67(Suppl 1):S12–6.

[5] Vorster HH. The link between poverty and malnutrition: a South African perspective. Health SA Gesondheid 2010;2010:15. (6p). Available at. http://www.hsag.co.za.

[6] Vorster HH, Badham JB, Venter CS. An introduction to the revised food-based dietary guidelines for South Africa. S Afr J Clin Nutr 2013;26(3):S1–64.

[7] Vorster HH, Kruger A, Margetts BM. The nutrition transition in Africa: can it be steered into a more positive direction? Nutrients 2011;3(4):429–41.

[8] Vorster HH, Wissing MP, Venter CS, et al. The impact of urbanization on physical, physiological and mental health of Africans in the North West Province of South Africa: the THUSA study. SA J Sci 2000;96:505–14.

CHAPTER

CANCER OF THE ESOPHAGUS IN SUB-SAHARAN AFRICA

7

Isidor Segal

G.I.T. Unit, Chris Hani Baragwanath Hospital and University of Witwatersrand, Soweto, South Africa

INTRODUCTION

Globally, IARC reported in 2002 that cancer of the esophagus is the 8th commonest cancer and the 6th commonest cause of cancer deaths worldwide. In developing countries it was the 5th commonest cancer, while in Sub-Saharan Africa it is the 4th commonest cancer (6%) [1,2].

The cancer is on the increase in Sub-Saharan Africa with uneven geographical distribution. It is the leading cancer in Southern and East Africa and rare in West Africa. There are certain high-risk areas in Kenya [2].

During preindependence of African countries, there were three marked peculiarities of esophageal cancer reported, namely, geographical distribution, vagaries of the sex distribution, and changing pattern of incidence with time [2]. It is noted that most of the data are incomplete and anecdotal in many of the Sub-Saharan countries. In 2002, it was only South Africa and Mauritius, which submitted the data of cancer mortality to WHO. The highest prevalence is in South and East Africa with *squamous* cancer the dominant cell type compared with *adenocarcinoma,* which is the commonest cell type of EC cancer in HICs.

In Ghana, the national trend of EC shows increasing incidence over the past 2 decades with alcohol and smoking the major risk factors (82% drank alcohol and smoked) [3]. As in other Sub-Saharan countries, squamous cell cancer is the dominant cell type (78% vs. adenocarcinoma (21.3%)). In Kenya, 98% of the cancers were of the squamous cell type and as in other Sub-Saharan countries late presentation was usual.

There are many reports from South Africa about esophageal cancer (EC). It was a rare disease in the South African Black population until the last few decades. Increases in incidence have occurred and at present it is the commonest cancer in Black men in many parts of South Africa. Studies of HICs populations have indicated a multiplicative effect of alcohol and tobacco in the development of EC with risk increasing more sharply with rising tobacco consumption [4]. This was confirmed in South African studies [5]. The highest incidence rates in Southern Africa are those in the south of the Transkei; 63 per 100,000 for males in 1981 and 65 for females.

Digestive Diseases in Sub-Saharan Africa. https://doi.org/10.1016/B978-0-12-815677-3.00007-4

SMOKING AND CE

The prevalence of smoking remains relatively low in SSA and smokers smoke fewer cigarettes than do their counterparts in Western and Asian countries. Significantly, the type of cigarettes smoked in many African countries is different to that of other regions of the world. In South Africa, a study of 200 CE patients and 391 controls showed that 75% of patients smoked cigarettes compared to 46% controls [5]. The survey also included questions about the use of hand-rolled cigarettes in which pipe tobacco is used wrapped in newspaper, brown paper, or telephone directory paper. Such cigarettes were often preferred because of their flavor and their slow burning quality. They were also cheaper than commercial cigarettes. Tobacco smoked either in pipes or hand-rolled cigarettes was found to be a major carcinogenic stimulus in black South African populations. The crude relative risk (RR) (RR: 1.07) indicated that there was no risk involved in smoking only commercial cigarettes [5]. However, in conjunction with other predictors, it becomes an important risk factor for esophageal cancer. Risk increased with increasing consumption for the smokers of both hand-rolled and commercial cigarettes.

RELATIONSHIP TO ALCOHOL AND DIET

Cook found evidence for a geographical and temporal association with the consumption of beer made from maize [6]. In many African countries, there has been a change in the type of ingredients used in making beer. An example of this is the pattern in South Africa where there has been a change from using mostly sorghum for beer-making to using mostly maize, with sorghum retained only as the fermenting agent. Bradshaw et al. [7] and Van Rensburg [8] investigated the dietary deficiencies that are common in areas of high incidence of traditional beer drinking and have suggested that low intakes of riboflavin, nicotinic acid, magnesium, and zinc are important underlying factors in development of the disease. However, in conjunction with other predictors, it becomes an important risk factor for CE patients.

It is noteworthy that consumption of traditional beer made from maize was found to be a major risk factor. This predisposition probably arises from a lifelong mild deficiency, usually subclinical, of several micronutrients as detailed here. It must be noted, however, that there is no consistent evidence of an effect of homemade brews and fruit as a CE risk in Africa [2].

CLINICAL CHARACTERISTICS

Dysphagia and weight loss are the main presenting symptoms. Other symptoms include abdominal pain, nausea, and anorexia. The cancer affects mainly men (M:F = 3:1) and mean age of presentation is 57 years for men and 51 years for women [5]. Late presentation is usual and prognosis is poor. In one series, 50% of patients had died after 7.8 months [9].

INVESTIGATIONS

Barium Swallow.

Upper gastrointestinal endoscopy.

Computer tomography and endoscopic ultrasound are used for cancer staging.

TREATMENT

Most patients in Sub-Saharan Africa present at a late stage with severe dysphagia and surgical resection is often the only option available.

In centers of excellence, premalignant lesions may be treated by endoscopic resection or ablation. Superficial cancer may be treated with endoscopic resection.

The optimal treatment for patients with localized esophageal cancer is surgical resection. Definitive chemoradiotherapy is an alternative treatment modality in nonsurgical candidates.

The palliation of patients with advanced nonresectable esophageal cancer involves treatment modalities such as endoscopic insertion of self-expandable metallic stents, endoscopic ablation, radiotherapy, and chemotherapy.

THE WAY FORWARD

There is a distinct lack of awareness that CE is the 4th commonest cancer in Sub-Saharan Africa, and the leading cancer in East and Southern Africa. The unified and integrated response to the AIDS pandemic can serve as a model for ameliorating the situation. This response led to multiple collateral benefits including a major increase in attention to and funding for global health issues [10]. It requires behavioral and structural changes at all levels from clinicians, public health officials, and leaders in the civil service to moderate this action. Evidence has shown that such an approach in conjunction with popular media support will assist in reversing, to a large extent, what is mainly a preventable cancer.

REFERENCES

[1] Kachala R. Systematic review: epidemiology of oesophageal cancer in sub-Saharan Africa. Malawi Med J 2010;22(3):65–70.

[2] Sitas F, Parkin M, Chirenje Z, et al. Cancer. In: Jamison DT, Makgoba MW, editors. Disease and mortality in Sub-Saharan Africa. 2nd ed. Washington, DC: World Bank; 2006.

[3] Tettey M, Edwin F, Aniteye E, et al. Changing epidemiology of oesophageal cancer in sub-Saharan Africa. Pan Afr Med J 2012;13:6.

[4] Day N, Munoz N. Esophagus. In: Schottenfield D, Fraumens J, editors. Cancer epidemiology and prevention. Philadelphia: WB Saunders; 2006. p. 596.

[5] Segal I, Walker ARP, Hamilton DG, Solomon A. Cancer of the oesophagous in the South African Black population. Trop Gastroenterol 1984;5:20.

[6] Cook P. Cancer of the oesophagous in Africa. Br J Cancer 1971;25:853.

[7] Bradshaw E, Schonland M. Smoking, drinking and oesophageal cancer in African males of Johannesburg, South Africa. Br J Cancer 1974;30:157.

[8] Van Rensburg SJ. Epidemiology and dietary evidence for specific nutritional predisposition to oesophageal cancer. J Natl Cancer Inst 1981;67:243.

[9] Walker ARP, Walker BF, Isaacson C, et al. Short duration of survival among South African Blacks with oesophageal cancer. S Afr Med J 1984;66:877.

[10] Piot P, Quinn TC. Response to the AIDS epidemic—a global health model. N Engl J Med 2013; 368(23):2210–8.

CHAPTER

8

DUODENAL ULCER AND DIET IN SUB-SAHARAN AFRICA

Frank Tovey

Department of Surgery, University College London, London, United Kingdom

INTRODUCTION

The ongoing research into the relationship between staple diets and the prevalence of duodenal ulcer began in India where the author was working as a surgeon in Mysore from 1951 to 1967 [1]. At that time, surgery was the accepted treatment for duodenal ulceration and was one of the major problems in male patients. The area to the north of Mysore was well irrigated and the staple crop was rice, whereas the region to the south was dry and a millet Ragi or Finger Millet (*Eleucine coracana*) or a pulse Horse Gram (*Dolichos biflorus*) were the staple crops. It became apparent that the majority of the patients were coming from the rice-growing areas. When this was realized, information was gathered over the next 10 years from all over India and a definite pattern emerged. The prevalence was high wherever rice or cassava were the staple foods and low in the regions where unrefined wheat or certain millets or pulses were the staple food [2].

From 1970 onwards until 1984, the author worked from the UK in conjunction with Denis Burkitt of the External Unit of the Medical Research Council. Denis Burkitt, of "Burkitt Lymphoma" fame, was interested in the effect of low or high dietary fiber content on the prevalence of diseases in developing countries, particularly in Africa, south of the Sahara. First of all, questionnaires were sent out to a large number of Mission Hospitals in Africa and also to India, Malaysia, and Indonesia, many of which sent monthly returns over a period of 3 years. A picture was built up from the correspondence, by personal interviews, by assessing reports from many sources and publications, from the opinion of clinical workers as to whether duodenal ulcer was a major or minor problem in the area where they worked, from radiological and necropsy records, and hospital admission figures comparing hospitals with similar facilities. In addition, the author accompanied Denis Burkitt on two visits to Africa to collect information, one to the West Coast to Togo, Ghana, the Ivory Coast, Sierra Leone, and Liberia, and one to East Africa to Kenya and Northern Tanzania. The author made two separate visits to Ethiopia and one to South Africa and Swaziland.

The period 1970–84 was an ideal time for collecting such information for two reasons. Firstly, the primary treatment for duodenal ulceration in developing countries was still surgical, and valuable information about local prevalence could be gathered from the number of surgical operations for chronic duodenal ulceration, or its complications, obtained from hospitals of similar size and with similar surgical facilities. Nowadays, with the widespread advent of medical treatment for duodenal ulceration, such information is no longer available. Secondly, in rural areas during this period staple diets were stable,

Digestive Diseases in Sub-Saharan Africa. https://doi.org/10.1016/B978-0-12-815677-3.00008-6

whereas nowadays, with better communications, there is much more variety in the diets in many rural areas and any relationship to a staple diet is no longer clear. Information from urban areas was less valuable because of more social classes and a wide variation in diets.

DUODENAL ULCER SOUTH OF THE SAHARA [3, 4]

High prevalence areas occurred along the West Coast in the hot humid coastal zones and adjacent rain forests, the highest prevalence being in the eastern area, in the Cameroons, Nigeria and on into Ghana, where the staple diet included yams, cassava, sweet potato, white flour, or rice. There was a very high prevalence in the Nile-Congo watershed. The prevalence was also high in the highlands of Ethiopia around Addis Ababa extending up to Gondar and Asmara where the staple diet was Teff, a small cereal grain cooked to make pancake-like enjerra.

The prevalence was low in the drier northern savannah regions of the West Coast, South Cameroon, Nigeria, Dahomey, Togoland, Ghana, Ivory Coast, Senegal, Sierra Leone, and Liberia, regions where millets, and more recently in parts of Northern Nigeria soya bean, were the staple foods. It was also low in northern Uganda, most of Zaire and Tanzania, Zambia, Malawi, Zimbabwe, Lesotho, and in southern Africa in Boputhasana, Venda, Transvaal, Transkei, Ciskei, and in Swaziland. These were areas where the staple foods were mostly millets or home-pounded maize and in some areas soya bean. It was rare also in the lowlands of Ethiopia where maize, millet, and wheat were grown.

The findings are shown in the two figures. Fig. 8.1 presents the overall results. Areas in which duodenal ulcer was common, occurred occasionally, or was uncommon are shown. Fig. 8.2 relates the number of duodenal ulcer cases in relation to total general hospital admissions excluding midwifery where the information was available. Three groups are shown as (1) less than one case per 1000 admissions; (2) one to 10 cases per 1000 admissions; and (3) over 10 cases per 1000 admissions.

PARTICULAR AREAS OF INTEREST

NILE-CONGO WATERSHED

The highest prevalence was in the moist region of the Nile-Congo watershed where duodenal ulcer formed the major part of all abdominal surgery. The area included Rwanda and Burundi, eastern Zaire around Lake Kivu, extreme western Tanzania adjacent to Burundi, and south-western Uganda. In Burundi, at Matana, a 58-bedded one-doctor hospital, 780 operations for peptic ulcer were done in 10 years (79% of all major surgery) and at Buye, a 76-bedded hospital, 404 operations for peptic ulcer were done in a period of 2 years, 8 months. The staple crops in this region were green bananas, sweet potato, cassava, and beans of the larger variety.

TANZANIA

The picture in Tanzania varied according to the climate and to staple crops.

There was a high incidence in the Wachaggas on the well-watered slopes of Mount Kilimanjaro and in the surrounding population. Their diet was essentially made up of green bananas with sour milk and occasional goat's meat. There was a rising incidence as one progressed to the west of Lake Victoria toward

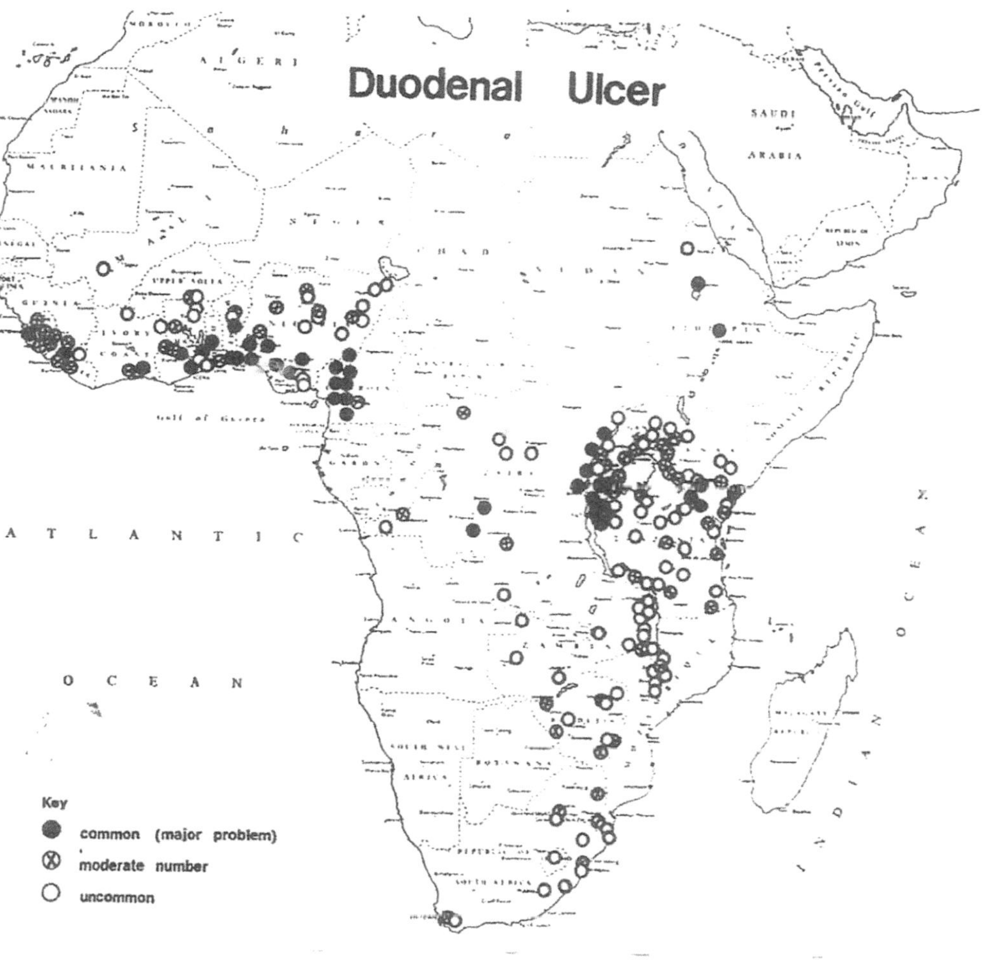

Map based on information from contacts and questionnaires

FIG. 8.1

Map based on information from contacts and questionnaires.

Burundi and Rwanda. The diet was largely green bananas. In the drier regions of Tanzania south of Mwanza, the incidence was low, duodenal ulceration being almost unknown in the dry millet-eating regions.

KENYA

The incidence was very low in the maize-eating areas of southern Kenya and also in the dry areas of the north where maize, millet, and locally taffi, a grass-like cereal, were the staple crops. There was a low-to-moderate incidence around Mount Kenya where there was a good rainfall, the diet on the west side being mostly maize and on the east side maize with barley, oats, and millets.

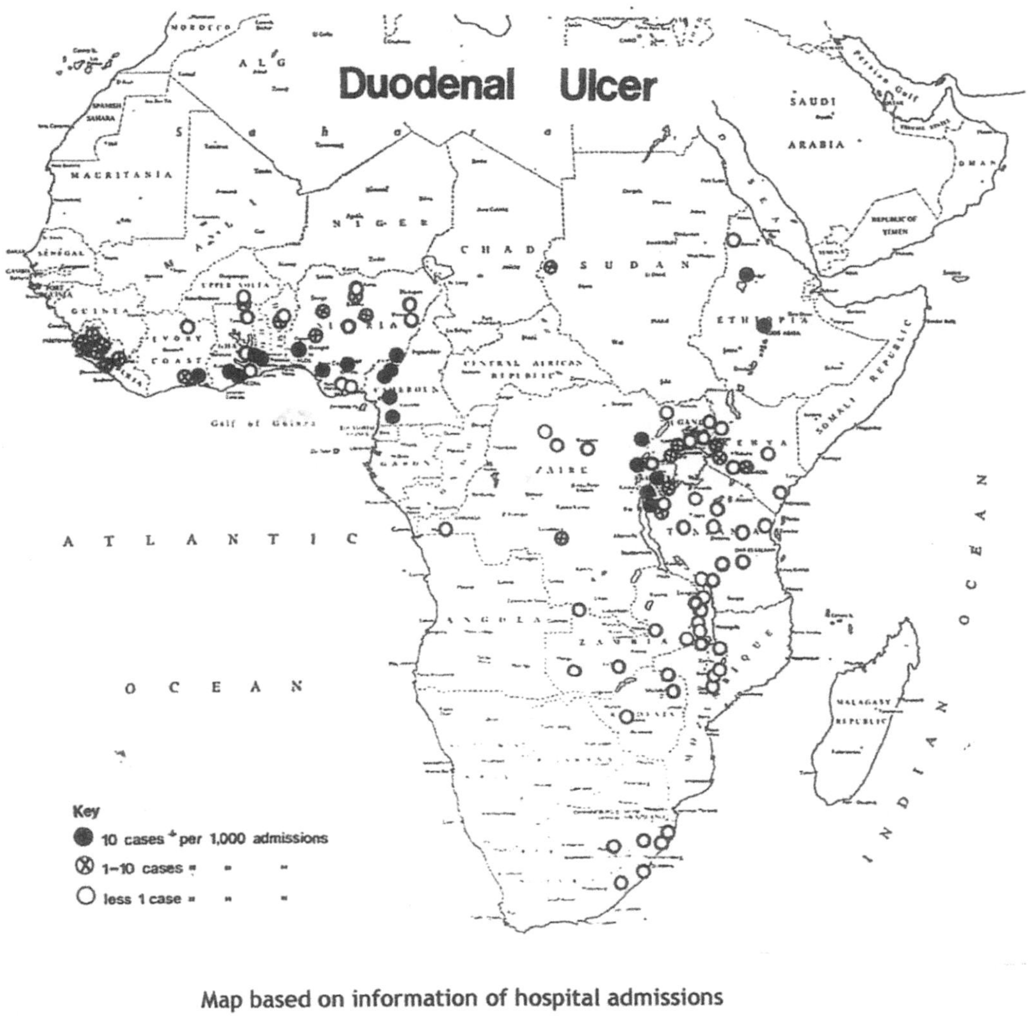

FIG. 8.2

Map based on information of hospital admissions.

THE MASAI AND OTHER MILK- AND BLOOD-EATING TRIBES

No duodenal ulceration was reported from the Masai, the Mangati nomads of Tanzania, or the nomads of Rendille in Northern Kenya. Their diet was mostly milk and its products, with occasional blood or meat. It was also unknown in the nomads in Gabbra in northern Kenya where the diet was principally camel's milk.

SOUTH AFRICA AND SOUTHERN AFRICA

In South Africa, the picture was mixed because of the mixed populations involved, many living in the towns, including Whites, Coloreds (largely Indian), and Blacks. Because of this, attention was focused on the black populations in rural areas.

Duodenal ulceration was almost unknown in the villages in the Black population of southern Africa, the Transkei, Ciskei, Lesotho, Boputhasana, Venda, and Transvaal where home-pounded maize was the principal staple food with millets in some areas, but was becoming common in the townships, particularly Soweto, where, because of economic pressures, it was easier to cook refined maize flour or rice, or to eat white bread.

In Cape Town, the available information related largely to the White population and the situation was similar to that in the West. The incidence in the Indian population resembled that in South India, except for the complications (see later). In Johannesburg, in the Baragwanath Hospital, the incidence in the Black population was slowly rising, especially in those who spoke English [5]. In the Non-European Hospital, the incidence in the Black Population had risen until 1976, since when it has remained steady [6]. In Durban, Professor M. Moshal of the King Edward VIII Hospital [7, 8] reported that duodenal ulcer incidence was rising in the Indian and Black population. In the Indian population and in those members of the colored population, it was more common in those who were on an Indian rice diet. The incidence was highest in the white population who were on a Western type of diet.

SWAZILAND

There was a low incidence throughout the country. The staple diet of the rural population was home-pounded maize with pumpkins and imported beans of the larger variety.

ETHIOPIA

Duodenal ulcer was common in the Highlands among the Amharic people whose staple diet was Teff, a grass-like cereal. It was relatively uncommon in the Lowlands where the staple crops varied according to altitude, temperature, and available land. In many areas, the staple foods were barley or wheat or maize or a millet (usually *E. coracana*), and in the South what was called False Banana (Ensette). Milk and its products featured in many of the low incidence areas.

DUODENAL ULCER CHARACTERISTICS AND EFFECT OF URBANIZATION

In the high prevalence areas of India, duodenal ulcer had distinctive characteristics. The commonest complication was pyloric stenosis, which often occurred early on in the history of the disease. At operation, one sometimes found a tumor-like inflammatory mass surrounding the first part of the duodenum and at other times severe fibrosis [2]. Hemorrhage and perforation were rare. The same situation was found in the rural high prevalence areas in Africa. It had been suggested that this might have been due to the difficulty of access to hospitals in rural areas, so that only chronic complaints were able to make the journey. However, the same picture was found in areas where the access was easy and many other acute conditions were being operated on in the hospitals.

In urban areas both in India and in Africa, the picture of duodenal ulceration was different; hemorrhage and perforation were common. This could be due to several factors in both countries. Firstly, there was a change in diet to using refined milled white rice, white wheat, or maize flour. This was not only due to availability but also due to the pressure of urban life, which left little time for home-pounding maize with its longer cooking times. For instance, in Soweto, most of the Black Workers would have to get up before dawn to get to work in Johannesburg in crowded trains and return after dark [9, 10]. Not only was the pressure of time but also the stresses of urban life, which were quite different from the stresses of life in a rural environment. These factors combined may well have led to a more acute type of ulceration resulting in the complications of hemorrhage and perforation. Stress is a known factor in duodenal ulceration. For instance, in the United Kingdom, during the War the incidence of perforation was high during air raids. A similar high incidence was reported during the riots in Soweto.

Another feature in the high prevalence areas of Africa was the almost complete absence of gastric ulceration in some areas with wide variation. In the United Kingdom, the ratio of duodenal ulcer to gastric ulcer in the 1950s was between 3:1 and 4:,1 and in the 1970s had fallen to between 2:1 and 3:1. In India, there was a wide range of 5:1 to 32:1, the mean from 18 reported series being 19:1 to 32:1. In Africa, the mean of 31 surgical reports was 33:1 (range being 4:1 to 156:1).

EVIDENCE RELATING TO THE EFFECT OF STAPLE DIETS ON DUODENAL ULCER PREVALENCE

The combined evidence suggests that where starchy diets such as yams, cocoyams, cassava, green bananas, refined white rice, wheat or maize flour are the staple food, the prevalence of duodenal ulcer is high, but where unrefined maize, millets, or soya bean form the staple foods the prevalence is low. In some high prevalence areas the larger pulses such as peas and beans feature, whereas in the low prevalence areas the smaller pulses feature in the diet.

HELICOBACTER PYLORI (*H. pylori*) INFECTION AND DUODENAL ULCER PREVALENCE

There is irrefutable evidence that eradication of *H. pylori* infection leads to healing of *H. pylori*–positive duodenal ulcers, but this does not necessarily mean that *H. pylori* infection is the primary cause of the ulceration. Infection with *H. pylori* is widespread, yet only a minority of infected persons develop duodenal ulceration. There is evidence that *H. pylori* infection has been prevalent for several centuries, yet duodenal ulceration only became common at the beginning of the twentieth century.

H. pylori infection could be a secondary and not a primary factor in duodenal ulceration. In Africa, the infection is almost ubiquitous affecting 70%–80% of the population often occurring in childhood. There was no relationship between the prevalence of *H. pylori* infection, its virulence, and the prevalence of duodenal ulceration. The prevalence of *H. pylori* infection was high in both high and low prevalence areas of duodenal ulceration [11–13]. This was called "The African Enigma" by Holcombe basing his information on Northern Nigeria [14, 15]. The African Enigma was disputed later by Agha and Graham [16] with information gathered from endoscopy findings from several centers in Africa saying that duodenal ulcer was common in all areas. On further examination of their evidence, however, it is apparent that the incidence of positive duodenal ulcer findings was much smaller in the low prevalence areas described here.

In support of information gathered from India and China, countries with a high prevalence of *H. pylori* infection showed no evidence of any correlation between the prevalence of duodenal ulceration and the prevalence of *H. pylori* infection or with virulence factors, the only correlation being that with the staple foods. In developed countries with a lower prevalence of *H. pylori* infection, a number of duodenal ulcers not linked with NSAIDs, especially early ulcers, are *H. pylori* negative. Duodenal ulceration can also recur despite eradication of *H. pylori* infection. Duodenal ulcers heal with proton pump inhibitors without the eradication of *H. pylori*. As many as 50% of acute perforated duodenal ulcers have been reported as *H. pylori* negative.

This suggests that *H. pylori* may be a secondary factor in duodenal ulceration [17–20] and the primary cause is high acid secretion [21] linked with reduced mucosal protection. When ulceration occurs, the presence of *H. pylori* could act like a secondary infection, interfering with the process of healing so that the ulcer becomes chronic [22]. Eradication of the *H. pylori* infection then allows the ulcer to heal.

PROTECTIVE FACTORS IN THE STAPLE DIET OF LOW PREVALENCE AREAS OF DUODENAL ULCERATION

ROLE OF DIETARY FIBER

A common feature of the staple foods of low prevalence areas, unrefined maize and wheat and certain millets as distinct from the starchy foods of the high prevalence areas, is that they have a high fiber content. Surgeon Captain T L Cleave [23, 24] and others thought that the protective activity of unrefined cereals and other foods was due to the fiber acting as a buffer against the acid secretion of the stomach. Gastric acid secretion studies using wheat bran however showed that although the bran initially acted as a buffer this action was offset by the action of the bran on the gastric antrum resulting in a big increase in acid secretion [25]. There were discrepancies also in the geographical prevalence of duodenal ulceration, such as a high prevalence in the highlands of Ethiopia where Teff (*Eragrastosis abyssinica*), rich in fiber was the staple food [3].

ROLE OF DIETARY LIPIDS

Research was then directed into the possibility of other factors related to fiber being present in the protective food stuffs. Water, ethanol and petroleum ether extracts from unrefined wheat, unrefined maize, soya, and certain protective millets and pulses were tested using rat peptic ulcer models and it was found that the lipid extract was consistently protective showing that the protective factors were present in the lipid fraction. Further tests done on fresh maize oil, whole wheat and wheat bran oil, soya bean oil, the oil from Finger Millet (ragi or *E. coracana*), and from the pulse Horse gram (*D. biflorus*) confirmed their protective activity [26–28].

THE NATURE AND ACTIVITY OF THE PROTECTIVE LIPIDS [29, 30]

The lipid extract from Horse gram (HGL) showed the greatest protective activity of all the lipids and was selected for further investigations using rat peptic ulcer models. It gave protection against ulceration in the rat pyloric ligation model, the rat ethanol, the rat aspirin, and the rat cysteamine model.

These models tested different modes of activity. The first three models showed protection against gastric ulceration. The pyloric ligation model involved 14 days' prefeeding with the protective lipid showing a systemic effect and the ethanol and aspirin model a single dose of the lipid just prior to giving the ethanol or aspirin showing a topical effect. In the pyloric ligation model the lipid gave protection against gastric acid, in the ethanol model it gave protection against a mucosal irritant, and in the aspirin model it showed protective activity against a nonsteroidal anti-inflammatory agent (NSAID). The cysteamine model showed topical protection following a single dose of the lipid against duodenal ulceration related to increased acid secretion.

The lipid was found to have not only a cytoprotective activity but was also found to promote ulcer healing. A daily dose promoted healing in both the ethanol and cysteamine models and also promoted healing of gastric ulceration produced by the topical application of acetic acid to the anterior wall of the rat's stomach.

RICE AND RICE BRAN OIL [31–33]

Fresh rice bran oil was protective but stored rice bran oil rapidly became ulcerogenic. This was because on milling, unlike wheat where the bran and germ are separated, rice bran and germ come off together on milling and the bruising activates the lipase present in the bran. The resulting lipolysis of the oil, which is present mainly in the germ, releases fatty acids. Then as a result of peroxidation, ketoaldehydes are produced from the fatty acids. Ketoaldehydes are ulcerogenic and the oil becomes ulcerogenic. Likewise the small oil content left in white rice after milling makes it become mildly ulcerogenic on storage.

MAIZE AND MAIZE OIL [31, 34]

Fresh maize oil obtained by 40–60 petroleum ether extraction from powdered maize was as protective as horse gram lipid in the rat ethanol model. Commercial maize oil obtained from a supermarket was not protective.

As in the case of wheat, the germ and the bran of maize are intimately bound together. Machine milling separates both the germ and the bran from the endosperm and from the resulting white maize flour. During home pounding, however, the bran and the germ remain mixed in with the resulting maize flour, which has to be eaten the same day because of the risk of rancidity due to lipolysis resulting from the mixing of the lipases with the oil present in the germ and bran.

Commercial maize oil is obtained from maize germ. The germ is heated to 121°C to destroy any lipase and prevent rancidity before the extraction of the oil. The loss of protective activity may be a result of this heat treatment.

INVESTIGATIONS INTO THE NATURE AND ACTIVITY OF COMPONENTS OF HGL [30]

Horse gram lipid was rich in phospholipids and sterols. Thin layer chromatography of HGL showed the presence of free fatty acids and triglycerides, phospholipids, sterol esters, and sterols. Different techniques were used to obtain fractions of HGL containing each of these components, which, in turn, were tested on different animal models for protective activity. The protective activity was shown to be lying in the phospholipid, the sterol, and sterol ester fractions.

PHOSPHOLIPIDS [30, 35]

The phospholipid fraction of HGL was protective in both the acute and chronic models. On high-performance liquid chromatography (HPLC) analysis, there were high peaks corresponding to phosphatidyl choline (lecithin) and ethanolamine (cephalin), and low peaks corresponding to lyso-phosphatidyl choline and ethanolamine, phosphatidyl serine, and phosphatidyl inositol. Based on the HPLC findings, preparations of two lecithins, dipalmitoyl phosphatidyl choline (DPPC) (saturated lecithin) and di-gamma linoleic acid phosphatidyl choline (unsaturated lecithin), and of one cephalin, dipalmitoyl phosphatidyl ethanolamine (saturated cephalin), were made in the laboratory for testing on the ratalcohol model and all gave significant protection.

Lichtenberger showed that phospholipids are cytoprotective, particularly Dipalmitoyl phosphatidyl choline (DPPC) and phosphatidyl choline. Phosphatidylcholine is present in plant sterols. DPPC is not found in plant material but is present in milk and animal fats. Phospholipids are integral components of cell membranes and of gastric and duodenal mucosal cells. They contribute to the viscosity of the surface layer of mucus and prevent intracellular acidification. The acyl side chains of unsaturated phospholipids supply unsaturated fatty acids for the synthesis of prostaglandins. They also have a surfactant action, forming a hydrophobic barrier to the effect of gastric acid. Romero and Lichtenberger reported that the protective effect of unsaturated phospholipids was enhanced by sterols (β-sitosterol and cholesterol), possibly due to enhancement of the packing of the unsaturated phospholipids in the cell membranes [36–39].

It has also been shown in rats and humans that either DPPC or phosphatidyl choline is protective against the injurious effects of nonsteroidal anti-inflammatory drugs (NSAIDs) [40–43].

STEROLS [30, 35, 44]

HPLC of the HGL fraction corresponding to commercial 60% β sitosterol was highly protective and showed high peaks corresponding to stigmasterol and β-sitosterol and smaller peaks corresponding to campesterol and Δ7-stigmasterol (Schottenol). On further analysis, five major subfractions were obtained and tested for protective activity using the ethanol model. Of these, two closely similar fractions were found showing highly significant activity ($P=.0006$). One of these was shown to contain β-sitosterol and stigmasterol and the other contained β-sitosterol and an unidentified isomer of β-sitosterol, which might be Y-sitosterol or Δ7-stigmasterol (Schottenol).

There are two reports about the ulceroprotective activity of sitosterols alone in animal models. These compare the protective activity of β-sitosterol with a glycoside. One is in Chinese [45] and reports on the antigastroulcerative activity of β-sitosterol-glycoside and its aglycone in rats using acetic acid and cold stress models, the effect being greater with the aglycone. The second paper [46] reports gastroprotective activity of β-sitosterol and sitosterol-3-0-b-glucoside in rat ethanol, aspirin, histamine, pyloric ligation, and 8-day glucose diet models, the sitosterolglycoside in this report being more effective than sitosterol.

Phytosterols regulate membrane fluidity and the activity of membrane-bound enzymes. They are active in reducing proton and sodium ion leaks from cell membranes and are reported to be anti-inflammatory, anti-pyretic, and to enhance immune response. Experimentally, certain phytosterols present in sponges and coral, but derived from plankton, have been shown to inhibit histamine release from peritoneal mast cells in rats. Phytosterols enhance the stability of phospholipid monolayers. They are reported to reduce the water permeability of phosphatidyl-choline bilayers [47–54].

STEROL ESTERS [30]

The sterols present in plant material always exist in combination with sterol esters and glycosides (sterolins). Sterol esters were protective in the rat pyloric ligation model after 14 days prefeeding with the sterol esters present in HGL, but were not protective when given in a single dose in the acute ethanol and cysteamine models. In contrast, sterols although protective in the acute models are not active in the long-term pyloric ligation model. The difference between sterol esters and sterols may be explained by the fact that sterols are poorly absorbed in the small intestine so their action in acute models on the gastroduodenal mucosa is topical, whereas sterol esters are absorbed and therefore will have a long-term systemic effect.

Three papers report on the protective activity of sterolins (sterol glycosides). Two papers [55, 56] report that unripe green banana (Musa Paradisiaca) in the Gangetic basin is protective against experimental ulceration in rats, using the aspirin model, between August and October when it contains a combination of four sitosterolins (sitoindoside I and II, sitosterol gentobioside, and sitosterol myo-inosityl-b-D-glucoside). The other paper [57], using mice-restraint models, reports the ulceroprotective activity of a mixture of steryl-b-D-glucosides.

REVIEW OF PROTECTIVE ACTIVITY

Phospholipids, sterols, and sterol esters in HGL all showed protective activity not only when given in combination but also when given individually, showing that each has its own particular contribution to the protection of gastroduodenal mucosa against peptic ulceration. They also give protection against ulceration due to NSAIDs.

HGL was also protective when given intramuscularly in the pyloric ligation and cysteamine models showing that it has a systemic as well as a topical effect. This systemic effect must be due however to the sterols and not the phospholipids present in the lipid, because phospholipids when given intramuscularly are absorbed locally and have no systemic effect. This confirms the protective activity of sterols when acting individually.

CONCLUSION

This information, which is largely based on information obtained 30 or more years ago, is still relevant to the present situation in Africa. In areas of lower rainfall where maize or millets are still the staple crops containing protective lipids, the prevalence of duodenal ulceration is likely to be low. Urbanization, increased mobility [9, 10, 58], the pressure of life, and machine milling [59] have resulted in the introduction of refined foods such as refined maize or wheat flour, milled white rice, and sugar, all of which are deprived of the lipids, which protect gastroduodenal mucosa, and this has been accompanied by an increasing prevalence of duodenal ulceration with its associated problems.

The protective activity of naturally occurring dietary phospholipids and phytosterols also has potential clinical value in the management of recurrent duodenal ulceration despite eradication of *H. pylori* infection and in the prevention of ulceration due to NSAIDs.

REFERENCES

[1] Tovey FI. Duodenal ulcer in Mysore. Trop Geogr Med 1972;24:107–17.

[2] Tovey FI. Peptic ulcer in India and Bangladesh. Gut 1979;20:329–47.

[3] Tovey FI, Tunstall M. Duodenal ulcer in black populations in Africa south of the Sahara. Gut 1975;16:564–76.

[4] Tovey FI. Duodenal ulcer. In: Burkitt D, Trowell H, Heaton K, editors. Dietary fibre, fibre-depleted foods and disease. London: Academic Press; 1985 [chapter 13].

[5] Segal I, Noormohamed AM, Ranchod S, Essoop AR, Oettle GJ. Duodenal and gastric ulcer in Soweto. S Afr Med J 1983;61:777–8.

[6] Bremner CG. Duodenal ulcer in the Johannesburg urban African. S Afr J Surg 1972;10:139–41.

[7] Robbs JV, Moshal MG. Duodenal ulceration in Indians and Blacks in Durban. S Afr Med J 1979;55:39–42.

[8] Moshal MG, Spitaels JM, Robbs JV, MacLeod IN, Good CJ. Eight-year experience with 3392 endoscopically proven duodenal ulcers in Durban, 1972–79. Gut 1981;22:327–31.

[9] Segal I, Dubb AA, Tim LO, Solomon A, Sottomayor MCCG, Zwane E. Duodenal ulcer and working class mobility in an African population in South Africa. BMJ 1978;469–72.

[10] Segal I, Unterhalter AM, Rosenbush H. Further observations on social factors associated with duodenal ulcer in Soweto. Soc Sci Med 1988;23:417–22.

[11] Holcombe C, Omotara BA, Eldridge J, Jones DM. *H. pylori*, the most common bacterial infection in Africa: a random serological study. Am J Gastroenterol 1992;87:28–30.

[12] Holcombe C, Eldridge J, Jones DM. Prevalence of antibody to *Helicobacter pylori* in children in northern Nigeria. Trans R Soc Trop Med Hyg 1993;87:19–212.

[13] Segal I, Ally R, Mitchell H. *Helicobacter pylori*—an African perspective. Q J Med 2001;94:561–5.

[14] Holcombe C. *Helicobacter pylori*: the African enigma. Gut 1992;33:429–31.

[15] Segal I, Ally R, Sitas F, Walker ARP. *Helicobacter pylori*: the African enigma. Gut 1998;43:300–1.

[16] Agha A, Graham DY. Evidence-based examination of the African enigma in relation to *Helicobacter pylori* infection. Scand J Gastroenterol 2005;40:523–9.

[17] Tovey FI, Hobsley M. Is *Helicobacter pylori* the primary cause of duodenal ulceration? J Gastroenterol Hepatol 1999;14:1053–6.

[18] Hobsley M, Tovey FI. *Helicobacter pylori*: the primary cause of duodenal ulceration or a secondary infection. World J Gastroenterol 2001;7:149–51.

[19] Hobsley M, Tovey FI. Precise role of *H. pylori* in duodenal ulceration. World J Gastroenterol 2006;(40):6413–9.

[20] Hobsley M, Tovey FI, Holton J. Controversies in the Helicobacter/duodenal ulcer story. Trans R Soc Trop Med Hyg 2008;102:6413–9.

[21] Hobsley M, Whitfield PF. The likelihood of a disease in relation to the magnitude of a risk factor. The example of duodenal ulcer. Theor Surg 1987;2:106–9.

[22] Pearce HR, Kalia N, Bardhan KD, Atherton JC, Brown NJ. Effects of *Helicobacter pylori* on epithelial cell proliferation and chemotaxis. Digestion 2004;69:201–10.

[23] Cleave TL. Peptic ulcer. Bristol: John Wright and Sons; 1962.

[24] Cleave TL. The saccharine disease. Bristol: Wright; 1974.

[25] Tovey FI. Aetiology of duodenal ulcer: an investigation into the buffering action and effect on pepsin of bran and unrefined carbohydrate foods. Postgrad Med J 1974;50:683–8.

[26] Jayaraj AP, Tovey FI, Clark CG. Possible dietary protective factors in relation to the distribution of duodenal ulcer in India. Gut 1980;21:1068–76.
[27] Tovey FI. Diet and duodenal ulcer. J Gastroenterol Hepatol 1994;9:177–85.
[28] Tovey FI. Staple diets and duodenal ulcer prevalence. Int Health 2009;1:124–32.
[29] Jayaraj AP, Tovey FI, Lewin M, Clark CG. Duodenal ulcer prevalence: experimental evidence for the possible role of dietary lipids. J Gastroenterol Hepatol 2000;15:610–6.
[30] Jayaraj AP, Tovey FI, Hobsley M. Duodenal ulcer: research into the nature of possible protective dietary lipids. Phytother Res 2003;17:391–8.
[31] Jayaraj AP, Tovey FI, Clark CG, Hobsley M. Peptic ulcer disease: dietary factors, its repair and relationship with *Helicobacter pylori* infection. J Gastroenterol Hepatol 2001;16:501–5.
[32] Gimeno FL, Bolant HB, Calvo BMA, Arino MJ, Palomer PEF, Amoros EJL, Lloris CJM, Narbona B. Antiulcerogenic properties of bran rice oil on rats. Rev Esp Enferm Dig 1989;75:225–30.
[33] Jayaraj AP, Rees KR, Tovey FI, White JS. A molecular basis for peptic ulceration due to diet. Br J Exp Pathol 1986;67:149–55.
[34] Tovey FI, Hobsley M, Segal I, Jayaraj AP. Duodenal ulcer in South Africa: home-pounded versus milled maize. J Gastroenterol Hepatol 2005;20:1008–11.
[35] Tovey FI, Bardhan KD, Hobsley M. Dietary phospholipids and sterols protective against peptic ulceration. Phytother Res 2013;27:1265–9.
[36] Murthy VLN, Sarosiek J, Siomiani BL. Effect of lipid and proteins on the viscosity of gastric mucus glycol-protein. Biophys Res Commun 1984;121:521–9.
[37] Lichtenberger LM, Graziani LA, Dial EJ, Butler BD, Hills BA. Role of surface-active phospholipids in gastric cytoprotection. Science 1983;219:1327–9.
[38] Hills BA, Butler BD, Lichtenberger LM. Gastric mucosal barrier: hydrophobic lining to the lumen of the stomach. Am J Phys 1983;219:311–26.
[39] Romero JJ, Lichtenberger LM. Sterol dependence of gastro-protective activity of unsaturated phospholipids. Dig Dis Sci 1990;35:1231–8.
[40] Lichtenberger LM, Wang ZM, Romero JJ, Ulloa C, Prez JC, Giraud MN, Barreto JC. Non-steroidal anti-inflammatory drugs (NSAIDs) associate with Zwitterionic phospholipids into the mechanism and reversal of NSAID-induced gastrointestinal injury. Nat Med 1995;1(2):154–8.
[41] Anand BS, Romero JJ, Sanduja K, Lichtenberger LM. Phospholipid association reduces the gastric mucosal toxicity of aspirin in human subjects. Am J Gastroenterol 1999;94:1818–22.
[42] Lanza FL, Marathi UK, Anand BS, Lichtenberger LM. Clinical trial: comparison of ibuprofen-phosphatidyl choline and ibuprofen on the gastrointestinal safety and analgesic efficacy in osteoarthritic patients. Aliment Pharmacol Ther 2008;28:431–42.
[43] Cryer B, Bhatt L, Lanza FL, Dong J, Lichtenberger LM. Low-dose aspirin induced ulceration is attenuated by Aspirin-Phosphatidylcholine: a randomized clinical trial. Am J Gastroenterol 2011;106:272–7.
[44] Tovey FI, Capanoglu D, Langley JG, Herniman JM, Bor S, Ozutemiz O, Hobsley M, Bardhan KD, Linclau B. Dietary phytosterols protective against peptic ulceration. Gastroenterol Res 2011;4:149–56.
[45] Xiao M, Yang Z, Jiu M, You J, Xiao R. The antigastroulcerative activity of beta-sitosterol-beta-D-glucoside and its aglycone in rats. Hua Xi Yi Ke Da Xue Xue Bao 1992;23:98–101.
[46] Navarette A, Trejo-Miranda JL, Reyes-Trejo L. Principles of root bark of *Hippocratea excelsa* (Hippocrataceae) with gastroprotective activity. J Ethnopharmacol 2002;79:383–8.
[47] Hennesy TM. Effect of membrane plant sterols on excitable cell function. Comp Biochem Physiol 1992;101:1–8.
[48] Haines TH. Do sterols reduce proton and sodium leaks through lipid bilayers? Prog Lipid Res 2001;40:299–324.
[49] Gupta MB, Nath R, Srivastava N, Shanker K, Kishor K, Bhargava KP. Anti-inflammatory and antipyretic activities of β-sitosterol. J Med Res 1980;39:157–63.

[50] Bouic PJ, Lamprecht JH. Plant sterols and sterolin: a review of their immune-modulating properties. Altern Med Rev 1999;4:170–7.
[51] Navarro A, De Las Heras B, Villar A. Anti-inflammatory and immunomodulating properties of a sterol fraction from *Sideritis foetens*. Chem Biol Pharm Bull 2001;24:470–3.
[52] Shoji N, Umeyama A, Takei M, Arihara S. Potent inhibition of histamine release; polyhydroxylated sterols from the Okinawan soft coral (*Sinularia abrupta*). J Pharm Sci 1994;83:761–2.
[53] Takei M, Burgoyne DL, Andersen RJ. Effect of contignasterol on histamine release induced by anti-immunoglobulin on rat peritoneal mast cells. J Pharm Sci 1994;83:1234–5.
[54] Hac-Wydro K, Wydro P, Jagoda A, Kapusta J. The study of the interaction between phytosterols and phospholipids in model membranes. Chem Phys Lipids 2007;150:22–34.
[55] Ghosal S, Saini KS. Sitoindosides 1 and 11. Two new anti-ulcerogenic steryl acyl-glycosides from Musa Paradisiaca. J Chem Res 1984;23:965–75.
[56] Ghosal S. Steryl glycosides and acyl steryl glycosides from *Musa paradisiaca*. Phytochemistry 1985;24:1807–10.
[57] Okyama E, Yamazaki M. The principles of Tetragonia tetragonoides having an anti-ulcerogenic activity isolation and identification of sterylglucoside mixture. Yakukagu Zasshi 1983;103:43–8.
[58] Segal I, Walker ARP, Vorster HH. Dietary factors associated with duodenal ulcer. BMJ 1991;32:456.
[59] Tovey FI, Hobsley M. Milling of wheat, maize: effects on fibre and lipid content and health. World J Gastroenterol 2004;10(12):1695–6.

CHAPTER

HELICOBACTER PYLORI IN SUB-SAHARAN AFRICA

9

Hazel M. Mitchell

Professor of Medical Microbiology, School of Biotechnology and Biomolecular Science, University of New South Wales, Sydney, Australia

INTRODUCTION

"In the world of modern medicine, it is rare that the understanding of a previously described clinical disease is so completely revolutionised that clinical textbooks must be rewritten. This however was the outcome of the isolation of Helicobacter pylori from the human stomach" [1]. The impact of this discovery was recognized by the award of the Nobel Prize for Physiology and Medicine to Robyn Warren and Barry Marshall in 2005. Thirty-five years following the initial isolation of *Helicobacter pylori* from the stomach of patients with gastritis and peptic ulcer disease [2], our understanding of the epidemiology and pathogenic mechanisms used by *H. pylori* to survive in the human stomach and to cause gastroduodenal disease has increased exponentially.

AFRICA: THE HOME OF *HELICOBACTER PYLORI*

Current evidence would suggest that the close association between humans and *H. pylori* originated in Africa. Analysis of the genetic makeup of *H. pylori* strains isolated from antral and corpus biopsies of 30 San individuals, a group of click-speaking hunter gatherers resident in South Africa, Northern Namibia, and Southern Angola, as well as a range of other sources, would suggest that *H. pylori* has been associated with humans for the last 88,000–116,000 years, with acquisition of *H. pylori* possibly occurring by a host jump from an unknown source [3]. In a study by Moodley et al., two discrete *H. pylori* superlineages were identified, one of which, until very recently, was predominantly associated with San (hpAfrica2) and large felines (*Hac*), while the second lineage is widespread throughout Africa (hpAfrica1, hpNEAfrica). Based on their findings, Moodley et al. suggested that this second lineage accompanied anatomically modern humans during their first "Out of Africa" migration, and that these lineages subsequently resulted in the Asian and Oceanic lineages, hpAsia2 hpAsia, and hpSahul [3]. Further, a study by Linz et al. has provided evidence that *H. pylori* left Eastern Africa in the company of its human host approximately 60,000 years ago [4].

Interestingly, a study by Olbermann et al. that compared the sequences of the complete *cytotoxin-associated gene A* (*cagA*) pathogenicity island (PAI) of 38 representative *H. pylori* isolates from subjects resident in all known *H. pylori* biogeographic populations has shown the gene content, structure, and function of the *cag* PAI to display a high level of global conservation, suggesting that a long shared evolutionary history exists between the *cag* PAI and the *H. pylori* core genome. Based on their findings,

Digestive Diseases in Sub-Saharan Africa. https://doi.org/10.1016/B978-0-12-815677-3.00009-8

Olbermann et al. proposed that the *cag* PAI was acquired prior to *H. pylori* leaving Africa with its human host [5].

Comparison of *H. pylori* strains isolated from Baka Pygmies, hunter-gatherers resident in the tropical rainforest in southeast Cameroon with *H. pylori* strains isolated from non-Baka subjects living in neighboring farming populations has shown Baka Pygmies to have a significantly lower prevalence of *H. pylori* (20.8%) than non-Baka subjects (80.2%) [6]. Interestingly, while comparison of the *H. pylori* strains isolated from Baka Pygmies with that of non-Baka agriculturalists showed the strains to be genetically similar, ancient *H. pylori* lineages were not identified in the Baka Pygmies. Based on these findings, Nell *et al.* hypothesized that either "the ancestors of the Baka Pygmies were initially *H. pylori*-free, or their ancestral bacteria were lost through past demographic fluctuations" [6].

THE EPIDEMIOLOGY OF *HELICOBACTER PYLORI* INFECTION IN SUB-SAHARAN AFRICA

Studies conducted in countries throughout Sub-Saharan Africa have shown the prevalence of *H. pylori* in healthy adults and children to remain high in the majority of countries within this region (Table 9.1), although a recent study by Etkuda et al. conducted in Lagos Nigeria over the period August 2008–2009 would suggest that the prevalence of *H. pylori* in children may be falling [7]. Consistent with the current literature, *H. pylori* infection in healthy subjects resident in sub-Saharan Africa is acquired early life and, in some countries, increases with age, while in others little change is observed with increasing age (Table 9.1).

Although over the last three decades a large number of studies worldwide have examined possible risk factors associated with the acquisition and transmission of *H. pylori*, as yet, the mode of transmission of this gastric pathogen remains poorly understood. Currently, person-to-person transmission of *H. pylori* is considered to be the major mode of transmission in both developed and developing countries, with transmission being postulated to occur predominantly via the fecal oral route, the gastro-oral route, or the oral-oral route [8]. As has been shown in other countries, in sub-Saharan Africa, transmission of *H. pylori* infection occurs within the family setting, with mothers in particular playing a key role in the transmission of *H. pylori* to their children [9–11]. For example, Aguemon et al. reported the risk of *H. pylori* infection in children in rural and urban Benin to be increased 13.6-fold when both parents were positive for *H. pylori* [95% OR=13.6 (3.63–51.22)], as compared with when either the mother or father were positive [(95% OR=5.3 (1.52–18.45) and 95% OR=2.7 (0.47–15.44)], respectively ($P<.0001$) [12].

Further, the number of subjects sleeping together has also been shown to increase the risk of *H. pylori* infection with Aguemon et al. reporting a 9-fold increased risk *of* infection when >3 persons shared a room (95% OR=9.82 (4.13–23.31), $P<.001$) [12]. Interestingly, a recent study by Didelot et al. investigating the genetic make-up of 97 *H. pylori* isolates cultured from 52 members of two South African families resident in Ogies, Mpumalanga, using 454 technology [13]. Based on comparison of the genomes of 45 family members in whom antral and body biopsies were available, Didelot et al. reported the transmission of *H. pylori* to be significantly more common in close relatives and in those living in the same house, a finding in keeping with earlier studies [13]. Interestingly, however, in 5 family members, the *H. pylori* isolates were so distantly related that Didelot et al. concluded that this most likely represented multiple infection, with infection likely to have occurred outwith the family

Table 9.1 The Prevalence of *Helicobacter pylori* Infection in Healthy Subjects in Sub-Saharan Africa

Country	Region	TEST	Age (yrs)	No. of Subjects	Prevalence *H. pylori* (%)	Refs
S. Africa	Bloemfontein	*H. pylori* IgG antibody	<2	104	13.5	[56]
			2–5	103	48.5	
			5–10	104	67.3	
			10–15	101	84.2	
S. Africa	Soweto	*H. pylori* IgG antibody	1 year	15	46.6	[57]
			12 years	?	100.0	
			18–79	75	81.3	
S. Africa	Limpopo Province	PCR	*All*	67	55.2	[58]
			3–5	10	80.0	
			6–9	4	25.0	
			10 15	58	55.2	
S. Africa	Eastern Cape Province	Stool *H. pylori* sandwich-enzyme immunoassay	*All*	356	86.8	[59]
			<12		75.9	
			25–47		100	
			≥60		100	
Cameroon	Baka Pygmies	Culture	Mean age (46.0 yrs)	77	20.8	[6]
	Non-Baka			101	80.2	
SW Cameroon	Buea and Limbe health districts	HPSA	*All*	176	52.27	[60]
			<3	32	37.50	
			3–6	106	50.00	
			7–10	38	71.05	
Benin	Cotonou	*H. pylori* IgG antibody	*All*	240 *Urban*	75.4	[12]
			2–5	21	76.2	
			6–9	38	84.2	
			10–13	43	72.1	
			14–15	11	81.8	
			16–24	23	82.6	
			25–34	37	67.6	
			35–44	53	75.5	
			45–54	9	77.8	
			55–74	5	40.0	
			All	206 *Rural*	72.3	
			2–5	23	60.9	
			6–9	37	73.0	
			10–13	17	58.8	
			14–15	9	66.7	
			16–24	21	76.2	
			25–34	40	80.0	
			35–44	32	84.4	
			45–54	10	70.0	
			55–74	17	58.8	

Continued

Table 9.1 The Prevalence of *Helicobacter pylori* Infection in Healthy Subjects in Sub-Saharan Africa—cont'd

Country	Region	TEST	Age (yrs)	No. of Subjects	Prevalence *H. pylori* (%)	Refs
Tanzania	North Mara District	*H. pylori* IgG antibody	*All*	787	92.0	[32]
			0–4	181	76.0	
			5–9	180	93.0	
			10–17	152	97.0	
			>18	268	99.0	
Nigeria	Lagos	*H. pylori* IgG antibody	*Overall*	100	59.0	[19]
			<14	47	35.6	
			≥15	53	64.4	
Nigeria	Lagos State	HPSA	*Overall*	185	72.4	[21]
			2–4	25	76.0	
			5–7	52	76.9	
			8–10	47	72.3	
			11–13	41	61.0	
			14–16	20	80.0	
Nigeria	Southern	*H. pylori* IgG antibody	Healthy adults	25	80.0	[61]
Southern Nigeria	Uyo	*H. pylori* IgG antibody	*Overall*	230	30.9	[7]
			<1	30	30.0	
			1–5	108	26.9	
			6–10	64	40.9	
			11–15	28	25.0	
Nigeria	Langat	HPSA-M	*Overall*	195	45.6	[62]
			0–6 mths	34	61.8	
			7–12 mths	46	41.3	
			13–18 mths	33	27.3	
			19–24 mths	39	41.0	
			25–30 mths	7	71.4	
			31–36 mths	36	52.8	

HPSA, H. pylori stool antigen; HPSA-M, H. pylori stool antigen monoclonal; PCR, polymerase chain reaction; IgG, Immunoglobulin G; Refs, References; mths, months; yrs, years.

setting [13]. In contrast, Delport et al. reported very high levels of gene diversity within black South African families resident in Ogies, Mpumulanga, the majority of individuals being shown to carry highly divergent genotypes, a finding the authors suggested indicated a large community reservoir of *H. pylori* genotypes as the source of infection [14]. Further, a study by Schwartz et al. reported no significant correlation between kinship and similarity of *H. pylori* sequences, nonunique isolates being less frequent in South African families, although individuals living in the same household were reported to have an increased probability of carrying the same nonunique isolates of *H. pylori,* independent of kinship [15]. Based on the high prevalence of *H. pylori* (86.8%) in fecal samples collected from healthy individuals resident in Eastern Cape Province of South Africa, Dube et al. have suggested that fecally contaminated water and food may represent an important source of *H. pylori* [16], a finding supported by Wubejig et al and Olefemi et al. [17, 18]. In contrast, Mynepalli reported no association to exist between drinking water source and *H. pylori* infection [19]. Given that the ability of *H. pylori* to survive in water remains unclear and that the detection of *H. pylori* by polymerase chain reaction (PCR) does not prove that the bacterium detected is viable, the importance of water as a source of *H. pylori* in Africa requires further investigation. While numerous studies worldwide have shown low socioeconomic status in childhood to be strongly associated with an increased prevalence of *H. pylori* infection, socioeconomic status is a wide-ranging criterion encompassing level of hygiene, sanitation, density of living, and educational opportunities, some, or all of which have been reported to influence the level of infection within a population [20]. In sub-Saharan Africa, socioeconomic status [21], educational level [22], plate sharing in the home [23], and premastication of food by the mothers prior to feeding their children [24] have all been shown to increase the risk of *H. pylori* infection. While several studies investigating the association between *H. pylori* and breastfeeding have reported a protective effect, this remains controversial [16, 60, 62].

THE AFRICAN ENIGMA

The discrepancy between the high prevalence of *H. pylori* in Africa and the low rate of *H. pylori*-related disease led Holcombe to describe this as "the African enigma" [25]. To investigate the African enigma, Kidd et al. conducted a retrospective literature review of all data published in Africa to determine if the prevalence of disease associated with *H. pylori* infection was similar to that in developed countries. Based on endoscopic examination of 3473 dyspeptic cases, Kidd et al. reported 26% of these cases to have DU (90% *H. pylori* positive), 7% gastric ulcer (GU((75% *H. pylori* positive), 28% atrophy (78% *H. pylori* positive), 14% metaplasia (78% *H. pylori* positive), and 3.4% gastric cancer (GC) (69% *H. pylori* positive). While this study would suggest that levels of duodenal ulcer (DU) and gastric ulcer (GU) in sub-Saharan Africa are similar to those observed in other countries, Kidd et al. concluded that it was difficult to establish a correlation between *H. pylori* and metaplasia and gastric cancer (GC) due the lack of good epidemiological data [26]. Evidence to support the African enigma has been reported by Segal *et al.* who described the situation in Soweto South Africa where, despite there being an extremely high prevalence of *H. pylori* (>90%), the incidence of GC is extremely low. Indeed, based on a review of the Chris Hani Baragwanath hospital records, which is the principal hospital serving Soweto, GC accounted for only 2.2% of all cancers diagnosed in the period 1948–64 and 2.8% in 1992 (an average of 40 cancers/year). Further, between 1990 and 1996, only 280 cases of GC were reported, which was extremely low given that the population of Soweto in 1995 was 3–4 million [27].

In an attempt to determine if evidence existed to support the African enigma, Agha and Graham undertook an evidence-based review of endoscopic data obtained from a range of African countries [28]. Based on Medline searches, 40 prospective endoscopic studies conducted in 17 countries within Africa in the period 1978 and 1994 were identified, which composed 20,531 patients (mean age: 31 to 53.1 years; male/female ratio=0.67:1–4.64:1). Determination of the *H. pylori*-related clinical outcomes in this cohort revealed 4326 patients (21.1%) were diagnosed with Duodenal ulcer (DU), 691 with Gastric Ulcer (GU) (3.4%) and 503 (2.4%) with gastric cancer (GC) Based on their results, Agha and Graham concluded that in Africa, "clinical outcomes associated with *H. pylori* infection were similar to those seen in industrialised countries," which they stated provided evidence that the African enigma never existed, but rather arose from reliance on anecdotal data, selection bias in populations with extremely limited access to health care and the relatively short life expectancy in Africa [28]. Importantly, however, these authors indicated that a number of questions remain unanswered regarding *H. pylori*-related disease in Africa, including whether a specific pattern of gastritis is associated with strains largely restricted to South Africa, and whether coinfection with parasites as suggested by Fox et al. might alter the expression of the infection, resulting in difference in the pattern and/or severity of gastritis, which may impact the prevalence of clinical disease [29, 30]. In line with the findings of Fox et al., Mitchell et al. demonstrated a difference in the T-cell response to *H. pylori* in symptomatic subjects from Soweto as compared with symptomatic subjects from Australia and Germany, Sowetan subjects being shown to mount a Th2-dominant *H. pylori*-specific immune response, while subjects from Australia and Germany mounted a Th1-dominant response. This study indicated for the first time that the host immune response to *H. pylori* infection in this Sowetan population differed to that in subjects from developed countries. While the effect of concurrent helminthic infection on the immune response to *H. pylori* in Sowetan subjects was not directly investigated, determination of total serum Immunoglobulin E (IgE) responses showed that 70% of symptomatic children and 58% of symptomatic adults to have total serum IgE levels well above the normal levels. Given that helminthic parasites are potent inducers of IgE, this finding indirectly supports a high prevalence of helminthic infection in the Sowetan population [31]. Similar findings have been reported by Mbulaiteye et al. who reported a Th2-dominant immune profile to be present in Tanzanian children and adults, this response being shown to be more frequent in children than adults (OR: 2.4, 95% CI: 1.3–4.4) [32]. Further, Campbell et al. have shown Gambian adults and children, as well as children from the United Kingdom (UK) to mount a predominantly Th2 response, while UK adults mounted a Th1 response. Based on their findings, Campbell et al concluded that Gambians, who are challenged with *H. pylori* infection in childhood, maintain the Th2 response throughout life, while UK adults appear to convert to a Th1 response. Based on their findings, the authors hypothesized that the apparent difference in prevalence of *H. pylori*-associated GC between West African and European populations may relate to a shift in polarity of T-helper-cell responses in chronic *H. pylori* infection [33].

HELICOBACTER PYLORI-RELATED DISEASE IN PATIENTS WITH DYSPEPSIA

Dyspepsia, defined as the presence of upper abdominal pain or discomfort with or without other upper gastrointestinal symptoms, such as nausea, belching, vomiting, is a common problem in Africa. While patients with such symptoms commonly visit their primary care physician, only a minority undergo endoscopic examination [34]. Based on studies conducted in sub-Saharan Africa over the last 13 years, the prevalence of *H. pylori* in dyspeptic patients varies considerably both between and within countries in this region (Table 9.2), ranging from 19.2% in Dar-es Salam, Tanzania, to 92.9% in Cameroon (Table 9.2). As discussed, such differences in prevalence are likely to relate to risk factors for infection such as level of hygiene, sanitation, density of living, and educational opportunities in these populations.

Table 9.2 The Prevalence of *Helicobacter pylori* Infection in Dyspeptic Patients Resident in Sub-Saharan Africa

Country	Region	Diagnostic Test	Age Group	Number of Patients With Dyspepsia	*H. pylori* Positive (%)	Refs
Cameroon	Douala Hospital	Culture	Mean age: 44.5 ± 15 yrs	77[a]	92.9	[63]
Ghana		RUT	15–87 yrs	130	75.4	[64]
Ghana	Kumasi	RUT	>15 yrs	300	58.72	[65]
S. Africa	Cape Town	Histology	NA	169	63.0	[6]
Ethiopia	Northwest	*H. pylori* IgG Ab antibody	Adults Median age: 38 yrs	209	72.7	[66]
				<20	60.0	
				21–30	70.2	
				31–40	79.1	
				41–50	77.3	
				>50	68.9	
Kenya	Nakuru	*H. pylori* IgG Ab	NA	138	71.0	[22]
Kenya	Nairobi	RUT and Histology	Adults Children	460	54.8	[67]
				236	73.0	
Kenya	Nairobi	RUT, Histology HPSAT, PCR	Adults	290	52.1[c]	[68]
			Gastritis		54.5	
			PUD		74.2	
Ethiopia	Bahir Da	EIA IgG Ab Immunoblotting	Mean age: 32.7 yrs	100	63–70[a]	[69]
Uganda	Kampala	Histology	Mean age: 43 ± 16 yrs	111	36.0	[70]
Northern Tanzania	Moshi	RUT	Median age: 48 yrs	N = 200	65.0	[72]
				Gastritis	72.0	
				Duodenitis	69.0	
				DU	89.0	
				GU	50.0	
				GC	58.0	
Tanzania	Dar-Es-salaam	*H. pylori* serology	Mean age: 44 yrs	135	19.3[a]	[73]

Continued

Table 9.2 The Prevalence of *Helicobacter pylori* Infection in Dyspeptic Patients Resident in Sub-Saharan Africa—cont'd

Country	Region	Diagnostic Test	Age Group	Number of Patients With Dyspepsia	*H. pylori* Positive (%)	Refs
Ethiopia	Addis Ababa	Culture	>15 years	300	69.0	[74]
		Urease			71.0	
		PCR-DGGE			91.0	
		Histopathology			81.0	
		Silver staining			75.0	
		HPSAT			81.0	
		Immunoblotting			80.0	
Ethiopia	Butajira	*H. pylori stool* Rapid Antigen Test Strip	Mean age: 48.6 ± 15.7 years	401	52.4	[75]
Ethiopia	Addis Ababa	*H. pylori* stool antigen and Helicobacter pylori rapid antibody	Adults	119	53.0	[76]
Ethiopia	Hawassa	*H. pylori* IgG A	>15 years	408	83.3	[77]
Nigeria	NA	RUT OR Histopathology	Adults	268	73.0	[78]
				Gastritis	76.0	
				Duodenitis	76.0	
				DU	76.0	
				GU	82.0	
				GC	50.0	
Nigeria	North-Eastern	Histology[a]	Mean age: 35.3 ± 12.7 yrs	125	80.0[a]	[79]
		H. pylori IgG antibody[b]			93.6[b]	
Northern Nigeria	Ado Ekiiti	*H. pylori* serology	18–50 yrs	103	22.3	[80]
Nigeria	Port Harcourt	^{14}C UBT	Mean age: 38.4 + 14.5 yrs	391	44.7	[34]
Nigeria	Ibadan	Histology[a]	NA	55	94.5	[81]
		H. pylori IgG Ab[b]				
Ghana	Accra	RUT	NA	242	74.8	[82]
Zimbabwe	University Zimbabwe	RUT and Histology	NA	95	79.0	[83]

Rwanda	Southern province	RUT-M	NA	All subjects:825	75.3	[84]
				Duodenitis: 67	6.97	
				DU: 148	5.40	
				GU: 26	2.70	
				GC: 14	1.45	
Ethiopia	Gondar	*H. pylori* IgG, IgM, IgA Ab[b]	NA	All: 1388	65.7	[85]
				0–20: 252	18.2	
				21–30: 466	33.6	
				31–40,256	18.4	
				41–50: 227	15.4	
				51–60: 113	8.1	
				>60: 74	5.3	
Nigeria	Maiduguri	2003–6	NA	All: 330	78.5	[86]
				Gastritis	89.1	
				DU	100	
				GU	61.9	
				GC	33.3	

[a] *91.11% of patients used and some were still using nonsteroidal antiinflammatory drugs (NSAIDs).*
[b] *Ab, antibody.*
[c] *Based on positivity in 2 or more tests.*
S. Africa, South Africa; Prov, province; RUT, rapid urease test; RUT-M, modified rapid urease test; ^{14}C UBT, ^{14}C urea breath test; PCR, polymerase chain reaction. HPSAT, H. pylori stool antigen test; HP-R IgG, H. pylori rapid IgG rapid antibody test; NA, not available.

DUODENAL ULCER

The distribution of duodenal ulcer disease in sub-Saharan Africa is not uniform, the distribution differing not only by country, but also within countries. Based on the data shown in Table 9.3, the highest rates of DU are found in Malawi, Cameroon, Kenya (Nairobi), and Nigeria, with a moderately high level being reported in Ghana. These findings are consistent with previous observations by Tovey et al. who reported Northern Tanzania, Ethiopia, the Nile/Congo watershed, the Cameroons, Nigeria, and Ghana to have the highest prevalence of DU [35].

GASTRIC ULCER

The distribution of gastric ulcer (GU) in sub-Saharan Africa is also variable, the highest prevalence rates being reported in Uganda (Mulago), Nigeria (Ibadan and Benin), Kenya (Nairobi 1993-4), and South Africa (Eastern Cape). While in some countries in sub-Saharan Africa GU is considerably less common than DU, this is not a consistent finding (Table 9.3). As shown in Table 9.3, the reported DU:GU ratios show considerable variation, ranging from 0.25:1 in Ibadan, Nigeria (Jemilohun et al., 2010) to 8.2:1 in Nigeria (Ndububa et al., 2001).

Interestingly, a retrospective endoscopic review of dyspeptic patients from 12 African countries by Kidd *et al.* reported 26.4% of patients to be diagnosed with DU while only 7% were diagnosed with GU. Of these, *H. pylori* was detected in 75% (Kidd et al., 1999).

GASTRIC CANCER

Based on Globocan 2012 statistics, an estimated 14.1 million new cancer cases and 8.2 million cancer deaths occurred worldwide in 2012. Of these, gastric cancer was the fifth most common malignancy, after cancers of the lung, breast, colorectum, and prostate, accounting for 6.8% of all cancer cases in adults worldwide (951,000 new cases) and the third most common cause of cancer death in both sexes worldwide (723,000 deaths), with incidence and death rates in males being higher in males, accounting for 456,000 cases and 362,000 deaths [36]. Data from the most recent Globocan report (2012) show the GC age standardized incidence and mortality rates in sub-Saharan Africa to be 3.9 and 3.7, respectively, with data on the incidence and mortality rates in the 3 major regions within sub-Saharan Africa showing the highest incidence and mortality rates to occur in Eastern Africa, and the lowest in Northern Africa (Table 9.4).

While in most developed countries cancer registries play an important role in assessing cancer incidence and mortality rates, in sub-Saharan Africa there are relatively few Cancer registries, these being Malawi (MCR), South Africa (PROMEC), Uganda, and Zimbabwe. Given this, reliable data regarding GC incidence and mortality rates in sub-Saharan Africa are relatively limited. However based on available data, estimated GC risks for countries within Sub-Saharan Africa have been determined (http://globocan.iarc.fr/Pages/Map.aspx), the estimated incidence rates in countries within sub-Saharan Africa are shown in Table 9.5. However, it should be noted that the reliability of the data varies considerably, the quality of the data being classified from A-G (Table 9.5).

Table 9.3 The Prevalence of Gastritis, Duodenal Ulcer (DU), Gastric Ulcer (GU), and Gastric Cancer (GC) in Patients With Dyspepsia Undergoing Endoscopy and the Association With *H. pylori* (1994–2015)

Country	Location	Time Frame	No. of Patients	Reason for Endoscopy	Endoscopy Findings							Refs
					Gastritis No. (%)	Duodenitis No. (%)	DU No. (%)	GU No. (%)	DU:GU Ratio	GC No. (%)		
Nigeria	Maiduguri	2003–2006	330	Dyspepsia	198 (60)	92 (27.9)	16 (4.8)	21 (6.4)	0.75:1	6 (1.7)		[86]
Kenya	Nairobi	1993–1994	120	Dyspepsia	NR	NR	35 (29.2)	11 (9.2)	3: 2:1	3 (6.7)		[87]
Kenya	Nairobi	2000–2001	71	Dyspepsia & Diabetes	48 (67.6)	17(25.7)	5 (5.7)	1 (1.4)	4.1:1	1 (1.4)		[88]
Northern Tanzania	Kilimanjaro	2010	208	Dyspepsia	127 (61.1)	32 (15.4)	38(18.3)	12(5.8)	3.2:1	14 (6.7)		[72]
Nigeria	NA	NR	834	Dyspepsia	NR	NR	NA (38.7)	NA (4.7)	8.2:1	52 (6.2)		[78]
Uganda	Mulago	2009–2010	167	Dyspepsia	NR	NR	13 (7.8)	18(10.8)	0.72:1	12 (7.2)		[89]
Nigeria	Benin	2006–2013	597	Dyspepsia	272 (45.6)	42 (7)	35 (5.9%)	51 (8.5)	0.7:1	10 (1.7)		[90]
Nigeria	Ibadan	2008–2009	86	Dyspepsia	52 (60.5)	NR	2 (2.3)	8 (9.3)	0.25:1	3 (3.5)		[91]
Kenya	Nairobi	NR	290	Dyspepsia	211 (73)	208 (71.7)	16 (5.5)	10 (3.4)	31(11.0)	2 (1.0)		[68]
Tanzania	Amana Municipal Hospital	2010–2011	393	Epigastric pain, Difficulty swallowing, Vomiting blood	152 (40.3)	NR	8 (2.1)	9 (2.4)	0.88:1	14 (3.7)		[92]
Uganda	Kampala	2014	111	Dyspepsia	82 (74.0)	NR	5 (4.5)	2 (2.7)	1.7:1	3 (2.7)		[70]
S. Africa	Eastern Cape Prov	2008	254	Dyspepsia	8 (3.1)	1 (0.3)	40 (15.7)	24 (9.4)	1.7:1	4 (1.5%)		[71]
Malawi	Zomba Hospital	2004–2006	441	Dyspepsia	44 (10.0)	NR	NR	NR	6 (1.4)	6 (1.4)		[93]
Cameroon	Banso	NR	93	Dyspepsia	82 (88.0)	NR	26 (28.0)	5 (5.4)	5.2:1	1 (1.0)		[94]
Africa	NA	1981–1997	4452	Dyspepsia	NR	NR	917(26.4)	160 (7%)	3.8:1	62 (3.4)		[26]
S. Africa	Ciskei & Transkei	NR	97	Dyspepsia	NR	NR	9 (9.3)	7 (7.2)	1.3:1	1 (1.0)		[95]
Sudan	Khartoum State	NR	200	Dyspepsia	164 (82.0)	26 (13.0)	NR	10 (5)	NR	NR		[96]
Ghana	Kumasi	2008	300	Dyspepsia	60 (20.0)	NR	32 (10.6)	13 (4.3)	2.5:1	7 (2.3)		[65]
Ghana	Kumasi	2006–2011	2542	Without alarm symptoms	NR	NR	75 (3.0)	116 (4.6)	0.65:1	37(1.5)		[97]

S. Africa, South Africa; NR, not reported; NA, not available in abstract; Refs, references.

Table 9.4 Gastric Cancer Age-Standardized Incidence and Mortality Rates ASR (W) per 100,000 Population, in Different Regions of Africa [36]

Region	Incidence ASR (W) per 100,000	Mortality ASR (W) per100000
Africa	3.8	3.5
Sub-Saharan Africa	3.9	3.7
Middle Africa	4.0	4.0
Eastern Africa	4.5	4.3
Northern	3.5	3.1

ASR, age standardized rate.

Based on the data presented in Table 9.3, it can be observed that in patients presenting with dyspepsia in sub-Saharan Africa the prevalence of GC varies considerably, the highest prevalence rates being reported in Uganda (7.2%), Northern Tanzania (6.7%), Nigeria (6.2%), and Kenya (6.7%), while the lowest rates were reported in Kenya (0.7%), Cameroon (1%), South Africa (1%), Malawi (1.4%), Kenya (1.4%), and South Africa (1.5%).

In a study to determine if pathological changes accompanying *H. pylori* infection might explain the varying prevalence of stomach cancer in Uganda, Wabinga investigated three ethnic groups resident in Uganda two of which have a high prevalence GC (Nyarwanda and Nkole) and Ganda, which has a low prevalence GC. This study showed that in the two high GC risk regions (Nkole and Nyarwanda) gastric atrophy was common, with *H. pylori* being detected in more severe forms of GA. Further, in these high-risk regions, intestinal metaplasia (IM) was also observed, with a number of these cases showing the presence of *H. pylori*. Based on these findings, Wabinga concluded that the major histological features relating stomach cancer to *H. pylori* were presence of the infection in IM and GA the two high GC areas (Nyarwanda and Nkole) [37].

H. PYLORI GENOTYPES AND DISEASE

The clinical outcome of long-term *H. pylori* infection has been reported to be associated with a range of bacterial virulence factors including the *cag* pathogenicity island (*cag*PAI), that encodes approximately 27–31 genes, including those encoding *cag*A and a type IV secretion system [38]. *cagA,* which is considered a marker of the *cag* pathogenicity island, is known to be polymorphic, having been shown to carry a different number of repeat sequences in the 3′ region of the *cagA* gene [39]. While in Western countries, individuals infected with *cagA*-positive strains of *H. pylori* are reported to be at an increased risk of developing peptic ulcer or gastric cancer than those infected with *cagA*-negative strains (39). A study conducted in South Africa by Kidd et al. has shown that >95% of isolates are *cag*A^{+}, and that 60% of these carry an intact contiguous *cag* PAI, which was present in 78% and 73% of PUD of GC isolates, respectively, which was significantly higher than that in isolates from those with gastritis alone (40%) ($P<.01$). Interestingly, in this study, deletions in the 5′ region of this genetic locus was reported to be associated with gastritis alone and the s2/m2*vacA* genotype [40]. The vacuolating cytotoxin A (encoded by *vacA* gene) has also been shown to be a major *H. pylori* virulence factor inducing a range of cellular events including vacuolation, membrane-channel formation, initiation of a proinflammatory

Table 9.5 Estimated Gastric Cancer Age-Standardized Incidence [ASR (W) per 100,000] and Reliability Scores for Countries Within Sub-Saharan Africa

Country	ASR (W) per 100,00	Reliability Score
South Africa	5.09	D
Mozambique	0.86	E
Botswana	0.91	D
Namibia	1.62	D
Angola	3.80	G
Democratic Republic of Congo	5.45	G
Tanzania	3.06	E
Zambia	4.40	E
Mali	9.80	E
Western Sahara	6.47	G
Mauritania	3.48	G
Nigeria	1.99	E
Niger	1.92	E
Burkino Faso	2.97	F
Ghana	2.88	F
Guinea	3.18	E
Senegal	6.24	G
Cameroon	2.36	E
Gabon	2.41	F
Seirra Leone	3.77	G
Liberia	3.67	G
Mozambique	0.86	E
Uganda	5.11	C
Cote D'Ivoire	4.36	F
South Sudan	4.96	G
Chad	2.01	G
Central African Republic	2.26	G
Kenya	9.50	E
Ethiopia	3.02	E
Somalia	6.31	G
Benin	3.72	F
Malawi	2.65	C
Zimbabwe	7.96	NA

Data obtained from website http://globocan.iarc.fr/Pages/Map.aspx; A. High-quality national data or high-quality regional (coverage >50%). B. High-quality regional (coverage between 10% and 50%). C. High-quality regional (coverage lower than 10%). D. National data (rates). E. Regional data (rates). F. Frequency data. G. No data [36].

response as well as cytochrome *c* release from mitochondria, which results in apoptosis and inhibition of T-cell activation and proliferation [39]. Determination of the genetic make-up of the vacA gene, which is present in all strains of *H. pylori*, has shown it to compose a mosaic structure and to contain allelic variations in both the signal (*s*) and mid (*m*) regions of the gene, these being *s*1/*s*2 and *m*1/*m*2. The biological activity of these vacA allelic types has been shown to differ with strains possessing the s1m1 genotype being associated with a higher prevalence of peptic ulcer and gastric cancer, than s2m2 [39]. A further *H. pylori* virulence factor that has been associated with disease outcome is the "induced by contact with epithelium" (*ice*A). The *iceA* gene has also been shown to have 2 allelic types, *iceA1* and *iceA2*, the *iceA1* allele being associated with peptic ulcer disease in some populations [41, 42].

H. PYLORI GENOTYPES IN SUB-SAHARAN AFRICA

A meta-analysis conducted by Sugimoto and Yamaoka investigating the relationship between *vacA* genotypes (*vacA* s, signal region variants; *vacA* m, middle region variants) and *H. pylori*-related disease development in 2285 Latin American strains and 434 African strains has reported 100% of patients with gastric cancer to carry the *vacA* s1 strain and >70% to carry the ml genotype. Further, the carriage of *vacA* s1 or m1 genotypes in African subjects was shown to increase the risk of peptic ulcer disease (8.69, 1.16–64.75 for s1) and gastric cancer (10.18, 2.36–43.84). In agreement with previous studies, strains carrying the s1 and m1 genotypes (62.1%) were frequently found to be *cag*A positive [43, 44].

A number of publications investigating the association between *H. pylori* genotypes and more severe disease outcomes have been published in the last 10 years, a number of which are shown in Table 9.6. For example, a study conducted in The Gambia by Secka et al. to determine the prevalence of *H. pylori* virulence factors in dyspeptic patients has shown cagA$^+$, *vac*As1, and *vac*Am1 genotypes to predominate. Interestingly, in their study, Secka et al subjects infected with a mixture of *cag*$^+$ and *cag*A$^-$ strains were shown to have a significantly lower disease burden as compared with those carrying *cagA*$^+$ or *cag*A$^-$ strains, a finding that led Secka *et al.* to suggest that harboring both *cagA*$^+$ or *cag*A strains may be protective against *H. pylori*-related disease [45]. Further in a study conducted in the Eastern Cape Province of South Africa by Tanih et al., the most common genotypes detected were *vacA* s1, *iceA2,* and *cagA+*. Although no significant correlation was observed between *cagA* status and PUD in this study the authors reported a higher prevalence of *cagA+* strains in PUD patients [46]. A further study conducted in Soweto South Africa has reported the majority of *H. pylori cagA*$^+$ strains to carry the *vac*As1 allele, while *cagA*$^-$ strains were found to predominantly carry the *vac*As$_2$ allele [47]. Interestingly, a study also conducted in South Africa by Kidd et al has reported a significantly higher prevalence of *ice*AI in patients with GC as compared with those with gastritis ($P<.001$). Further, the vacAs1/iceA1 genotype was found to be significantly higher in isolates from PUD patients as compared those with gastritis alone ($P<.001$) [48].

ANTIBIOTIC RESISTANCE

Antibiotic resistance is an escalating problem worldwide, particularly in developing countries due to misuse of antibiotics as well as the availability of over-the-counter antibiotics. Like many bacterial pathogens, *H. pylori* has developed resistance to a range of antibiotics previously shown to achieve eradication. Based on the most recent consensus statement, a combination of two antibiotics, either

Table 9.6 The Prevalence of *H. pylori cag*A and *vac*A and *ice*A Genotypes in Sub-Saharan Africa

		vacA Genotypes (%)									*iceA* Genotypes			
		vacA	*vacA*	*vacA*	*vacA*	*vacA*	*vacA*	*vacA*	*vacA*	*vacA*	*iceA*	*iceA*	*iceA*	
Country	***cagA***	**s1**	**s2**	**s1/s2**	**m1**	**m2**	**m1/m2**	**s1m1**	**s1m2**	**s2m2**	**1**	**2**	**2-Jan**	**Refs**
S. Africa	90	58	26	16	8	50	36	8	20	20	2	58	26	[46]
Ethiopia	79	88.5	11.5	0	57	43	0	48	28	9	NT	NT	NT	[98]
Africa[a]	89	83.9	16.1	0	56.7	43.3	0	52.5	31.5	14.7	NT	NT	NT	[43]
S.Africa	83.3	NT	NT	NT	NT	NT	NT	66.6	16.6	16.6	NT	NT	NT	[99]
Nigeria	100	NT	NT	NT		NT	NT	87.5	12.5	0	NT	NT	NT	
S. Africa														
DU	81.8	90.9	9.1	0	63.6	36.4	0	54.5	36.4	0	NT	NT	NT	[100]
No ulcer	40	100	0	0	100	0	0	100	0	0	NT	NT	NT	
S. Africa														
Overall	95	80	20	0	68	32	0							[101]
Gastritis	66	65.7	34.4	0	57	42	0	NT	NT	NT	NT	NT	NT	
PUD	100	100	0	0	71	29	0							
GC	100	100	0	0	100	0	0							
S. Africa														
Gastritis	100	ID	ID	ID	ID	ID	ID	ID	ID	ID	23	86	37	[48]
PUD	100	ID	ID	ID	ID	ID	ID	ID	ID	ID	41	77		
GC	100	ID	ID	ID	ID	ID	ID	ID	ID	ID	67	47		
Kenya														
Overall	44.6	42.7	42	0	63.3	63.3	0	NT	NT	NT	NT	NT	NT	[68]
Gastritis	40	42.6	42.6	0	68.7	68.7	0	NT	NT	NT	NT	NT	NT	
PUD	65.2	30.4	30.4	0	39.1	39.1	0	NT	NT	NT	NT	NT	NT	
Gambia	59	74	22	0.78	44.1	32.3	17.3	NT	NT	NT	14.1	29.1	2.5	[102]

Continued

Table 9.6 The Prevalence of *H. pylori cag*A and *vac*A and *ice*A Genotypes in Sub-Saharan Africa—cont'd

		vacA Genotypes (%)									*iceA* Genotypes			
		*vac*A	*vac*A	*vac*A	*vac*A	*vac*A	*vac*A	*vac*A	*vac*A	*vac*A	*ice*A	*ice*A	*ice*A	
Country	***cagA***	**s1**	**s2**	**s1/s2**	**m1**	**m2**	**m1/m2**	**s1m1**	**s1m2**	**s2m2**	**1**	**2**	**2-Jan**	**Refs**
Nigeria														
All	92.7	98	2	ID	ID	ID	ID	24.4	73.2	4.9	90.2	7.3	2.5	[103]
DU	94.7			ID	ID	ID	ID	26.3	73.7	4.5	94.7	5.3	0	
Dyspepsia	90.9			ID	ID	ID	ID	22.7	72.7	2.5	86.4	9.1	4.5	
Gambia	61.2	76.9	19	0.8	45.5	29.8	18.2	43.8	14.9	14.9	NT	NT	NT	[45]
Senegal														
GC	93.3	100	0	0	80	80	0	ID	ID	ID	93.3	6.7	0	[104]
Other	70	22.2	100	0	6.8	93.2	0	ID	ID	ID	75.5	24.5	0	

[a]*Data based on the combined results of 6 studies conducted in Egypt (1 study), Ethiopia (1 study), Nigeria (2 studies), South Africa (3 studies).*
S. Africa, South Africa; Data based on a meta-analysis; NT, not tested; EX, excluded; ID, insufficient data to determine values; cagA, cytotoxin-associated gene A; vacA, vacuolating cytotoxin; iceA, induced by contact with epithelium gene A; Refs, references.

amoxicillin or metronidazole (based on local resistance rates), clarithromycin, and a proton pump inhibitor are recommended as first-line therapy for the treatment of *H. pylori* [49]. In a number of countries within Sub-Saharan Africa, extremely high rates of resistance have been reported to these and other antibiotics used in the treatment to *H. pylori*, although significant variation in resistance rates exist in countries within this region (Table 9.7). Such variation has been related to the level, use, and free access to antimicrobial agents in some countries within Sub-Saharan Africa [50]. For example, in Nigeria, the common use of ampicillin, tetracycline, metronidazole, and erythromycin has led to high resistance rates to these antibiotics (Table 9.7). In all countries within sub-Saharan Africa, *H. pylori* has been shown to have high levels resistance to metronidazole, ranging from 40% to 100% (Table 9.7). Further, in a number of countries within the region, resistance rates to amoxicillin and tetracycline are also high (Table 9.7). Interestingly, resistance rates in sub-Saharan Africa show a similar trend to those reported in a recent meta-analysis conducted by Camargo et al., which examined *H. pylori* resistance rates in Latin America. This showed specific resistance rates to amoxicillin to range from 0% to 39%, for tetracycline from 0% to 86%, for clarithromycin from 0% to 60%, for metronidazole from 12.5% to 95%, and for fluoroquinolones 4% to 37% [51]. Given the increasing resistance rates of *H. pylori* to currently used antibiotics, a number of alternative approaches to treatment have been investigated. An excellent review of this topic has been published by *Samie* et al. [52]. Interestingly despite high resistance rates, a recent study by Onyekwere has reported some positive findings in relation to the treatment of *H. pylori* infection. In this study conducted in subjects resident in Lagos, Nigeria, a combination of amoxicillin, clarithromycin, and rabeprazole was found to have high efficacy, the average eradication rate of 87.2% that was not influenced by the gender or age of the subjects. Interestingly, no significant advantage was observed in the efficacy of 10-day treatment duration as compared with 7 days, no statistically significant differences being observed in eradication rates between those treated for 7 d (88.9%) and those treated for 10 d (86.2%) [53].

H. PYLORI AND ASSOCIATION WITH OTHER DISEASES IN AFRICA

A recent meta-analysis by Li et al. examining the association between *H. pylori* and hyperemesis gravidarum (HG) has shown a significant association to exist between *H. pylori* and HP, the *H. pylori* infection rate being significantly higher in pregnant women with HG (OR = 3.34, 95% CI: 2.32–4.81, $P < .001$) [54]. Interestingly, subgroup analysis indicated that *H. pylori* infection was a significant risk factor for HG in Asia, Africa, and Oceania, the risk being particularly high in Africa, where those infected with *H. pylori* were shown to have a >12 times increased risk of HG (OR = 12.38, 95% CI: 7.12–21.54, $P < .001$). Based on these findings, the authors concluded that *H. pylori* should be considered one of the risk factors of HG, especially in the developing countries [54]. Further, Malande et al. have reported three case histories of *H. pylori* gastritis in Kenyan indigenous black African female infants aged 4, 6, and 7 months. On presentation, the three infants displayed symptoms ranging from fever, refusal to feed, diarrhea, restlessness, vomiting, and irritability. Based on laboratory investigations, all three infants were infected with *H. pylori*. Following treatment with a combination of IV/oral omeprazole, oral clarithromycin, and IV/oral metronidazole for 14 days all 3 infants improved, complete resolution of symptoms and negative stool antigen tests being observed at completion of therapy [55].

Table 9.7 Antibiotic Resistance Rates in Countries Within Sub-Saharan Africa

Country/ Time Frame	No. *H. pylori* Isolates (Source)	Resistance Rates (%) to Antibiotics Commonly Used to Treat *H. pylori* Infection in Subjects From Different Regions of Sub-Saharan Africa								Refs
Antibiotics		AMO	TET	METRO	CLARI/ ERYTH[a]	LEVO	CIPRO	RIFAMP	GENT/ KANA[#]	
Cameroon	132 Adults	85.6	43.9	93.2	44.7	NT	NT	NT	NT	[63]
Ethiopia	50 Adults	6.0	0.0	76.0	0.0 0.0[a]	NT	NT	NT	NT	[105]
Gambia	58 Adults	0.0	0.0	76.0	0.0	NT	NT	NT	NT	[106]
	6 Infants	0.0	0.0	0.0	0.0	NT	NT	NT	NT	
Kenya	65 Adults and children	0.0	NT	4.6	0	NT	NT	NT	NT	[67]
Senegal	108 (Adults)	0.0	0.0	85.0	1.0	15	NT	NT	NT	[107]
Senegal	40 (Adults)	40	NA	90.0	NT	NT	0	NT	NT	[108]
Nigeria Jos	55 adults	0.0	11.0	40.0	12.7	NT	NT	NT	NT	[109]
W. Nigeria	56 Adults	100.0	100.0	100.0	100.0[a]	NT	0.0	NT	75.0	[110]
SW. Nigeria	32 Adults	100	87.1	100	100	15.6	NA	93.54	NT	[50]
S. Africa	200 Adults	2.5	32.5	95.5	44.5	NT	0	NT	27.5	[111]
S. Africa[a]	78 Adults[a]	NT	NT	NT	15.4	10.3[a]	NT	NT	NT	[112]
	50 Males				6.0	6.0	NT	NT	NT	
	28 Females				31.2	17.9	NT	NT	NT	

[a]*No previous treatment with clarithromycin or quinolones.*
S. Africa, South Africa; AMO, amoxycillin; TET, tetracycline; CLARI, clarythromicin; ERYTH, erythromycin; LEVO, levofloxacin; RIFAMP, rifampicin; GENT, gentamicin; KANA, kanamycin; NT, not tested; Refs, References.

CONCLUSION

Studies conducted in sub-Saharan Africa indicate that the prevalence of *H. pylori* infection remains high in the majority of countries within this region and that infection occurs early in life, acquisition being predominantly associated with transmission within the family. In contrast to previous views, which suggested that duodenal ulcer disease and gastric ulcer were rare in sub-Saharan Africa, more recent data demonstrate that the prevalence of duodenal ulcer disease and gastric ulcer in sub-Saharan Africa is in general similar to that reported in other countries, and that *H. pylori* is strongly associated with these disease outcomes. In contrast, the prevalence of gastric cancer varies within this region with some countries reporting high *H. pylori* prevalence rates but demonstrating low prevalence rates of gastric cancer, a finding that may relate to coinfection with parasites or host genetic factors, the latter issue having received limited attention in this region. In many countries within sub-Saharan Africa, the resistance rates of *H. pylori* to currently recommended antibiotic regimens are very high, which has resulted in failure to eradicate the infection. Given this, research into alternate therapies for the treatment of *H. pylori* infection are currently being investigated in this region.

REFERENCES

[1] Mitchell HM. The epidemiology of *Helicobacter pylori* infection. In: Westblom TU, et al., editors. Gastroduodenal disease and *Helicobacter pylori*. Berlin, Heidelberg: Springer1-Verlag; 1999.

[2] Marshall BJ, Royce H, Annear DI, et al. Original isolation of *Campylobacter pyloridis* from human gastric mucosa. Microbios Lett 1984;25:83–8.

[3] Moodley Y, Linz B, Bond RP, et al. Age of the association between *Helicobacter pylori* and man. PLoS Pathog 2012;8:e1002693.

[4] Linz B, Balloux F, Moodley Y, et al. An African origin for the intimate association between humans and *Helicobacter pylori*. Nature 2007;445:915–8.

[5] Olbermann P, Josenhans C, Moodley Y, et al. A global overview of the genetic and functional diversity in the *Helicobacter pylori cag* pathogenicity island. PLoS Genet 2010;6:e1001069.

[6] Nell S, Eibach D, Montano V, et al. Recent acquisition of *Helicobacter pylori* by Baka pygmies. PLoS Genet 2013;9:e1003775.

[7] Etukudo OM, Ikpeme EE, Ekanem EE. Seroepidemiology of *Helicobacter pylori* infection among children seen in a tertiary hospital in Uyo, southern Nigeria. Pan Afr Med J 2012;12:39.

[8] Mitchell HM. The epidemiology of *Helicobacter pylori*. Curr Top Microbiol Immunol 1999;241:11–30.

[9] Drumm B, Perez PGI, Blaser MJ, Sherman PM. Intrafamilial clustering of *Helicobacter pylori* infection. N Engl J Med 1990;322:359–63.

[10] Malaty HMKT, Tanaka E, Ota H, Kiyosawa K, Graham DY, Katsuyama T. Transmissin pathways of *Helicobacter pylori* infection: results from a 9-year birth cohort study in Japan. J Clin Microbiol 2000;38:1971–3.

[11] Mitchell HM, Bohane T, Hawkes RA, et al. *Helicobacter pylori* infection within families. Zentralbl Bakteriol 1993;280:128–36.

[12] Aguemon B, Struelens M, Massougbodji A, et al. Prevalence and risk-factors for *Helicobacter pylori* infection in urban and rural Beninese populations. Clin Microbiol Infect 2005;11:611–7.

[13] Didelot X, Nell S, Yang I, et al. Genomic evolution and transmission of *Helicobacter pylori* in two South African families. Proc Natl Acad Sci 2013;110:13880–5.

[14] Delport W, Cunningham M, Olivier B, Preisig O, van der Merwe SW. A population genetics pedigree perspective on the transmission of *H. pylori*. Genetics 2006;174:2107–18.

[15] Schwarz S, Morelli G, Kusecek B, et al. Horizontal versus familial transmission of *Helicobacter pylori*. PLoS Pathog 2008;4:e1000180.

[16] Dube C, Tanih NF, Ndip RN. *H. pylori* in water sources: a global environmental health concern. Rev Environ Health 2009;24:1–14.

[17] Olufemi FO, Remi Q, Akinduti PA, et al. Potential risk factors and prevalence of infection of *Helicobacter pylori* in Nigeria. J Sci Res Rep 2015;7:42–8.

[18] Wubejig A, Mulugeta K, Bayeh A. Prevalence and risk factors of *H. pylori* from dyspeptic patients in Northwest Ethiopia: a hospital based cross-sectional study. Asian Pac J Cancer Prev 2014;15:4459–63.

[19] Mynepallil SKC, Maureen O, Adejumo M. Prevalence of *Helicobacter pylori* and hygiene practices among public secondary school students in Ikeja local government area, Lagos, Nigeria. Health 2014;6:250–8.

[20] Mitchell HM. Epidemiology of infection. In: HLT M, Mendz GL, Hazell SL, editors. *Helicobacter pylori*: physiology and genetics. Washington, DC: ASM Press; 2001. Chapter 2.

[21] Olufemi F, Remi Q, Shittu O, et al. *Helicobacter pylori* stool antigen in asymptomatic children in Lagos State, Nigeria. Int J Trop Dis Health 2015;7:156–62.

[22] Shmuely H, Obure S, Passaro DJ, et al. Dyspepsia symptoms and *Helicobacter pylori* infection, Nakuru, Kenya. Emerg Infect Dis 2003;9:1103–7.

[23] Nabwera HM, Nguyen-Van-Tam JS, Logan RF, Logan RP. Prevalence of *Helicobacter pylori* infection in Kenyan children age 3-15 years and risk factors of infection. Eur J Gastroenterol Hepatol 2000;12:483–7.

[24] Albenque M, Tall F, Dabis F, Mégraud F. Epidemiological study of *Helicobacter pylori* transmission from mother to child. Rev Esp Enferm Dig 1990;78:48.

[25] Holcombe C. *Helicobacter pylori*: the African enigma. Gut 1992;33:429–31.

[26] Kidd M, Louw JA, Marks IN. *Helicobacter pylori* in Africa: observations on an 'enigma within an enigma'. J Gastroenterol Hepatol 1999;14:851–8.

[27] Segal I, Ally R, Mitchell H. *Helicobacter pylori*—an African perspective. Q J Med 2001;94:561–5.

[28] Agha A, Graham DY. Evidence-based examination of the African enigma in relation to *Helicobacter pylori* infection. Scand J Gastroenterol 2005;40:523–9.

[29] Fox JG, Wang TC, Nagler-Anderson C. The African enigma: the parasite's perspective. Gut 2001;49:156–7.

[30] Gordon D. Solving the African enigma: parasites may have their place. Gastroenterology 2000;119:61.

[31] Mitchell HM, Ally R, Wadee A, et al. Major differences in the IgG subclass response to *Helicobacter pylori* in the first and third worlds. Scand J Gastroenterol 2002;37:517–22.

[32] Mbulaiteye SM, Gold BD, Pfeiffer RM, et al. *H. pylori*-infection and antibody immune response in a rural Tanzanian population. Infect Agent Cancer 2006;1:3. https://doi.org/10.1186/1750-9378-1-3.

[33] Campbell DI, Pearce MS, Parker L, Thomas JE, Sullivan PB, Dale A. Immunoglobulin G subclass responses to *Helicobacter pylori* vary with age in populations with different levels of risk of gastric carcinoma. Clin Diagn Lab Immunol 2004;11:631–3.

[34] Bock-Oruma AA, Dienye PO, Akani AB, et al. Prevalence of *Helicobacter pylori*-related dyspepsia using a near-patient testing method in Port Harcourt, Nigeria. TJFMPC 2013;7:29–34.

[35] Tovey FI, Tunstall M. Duodenal ulcer in black populations in Africa south of the Sahara. Gut 1975;16:564–76.

[36] Ferlay J, Soerjomataram I, Dikshit R, Eser S, Mathers C, Rebelo M, Parkin DM, Forman D, Bray F. Cancer incidence and mortality worldwide: sources, methods and major patterns in GLOBOCAN 2012. Int J Cancer 2015;136:E359–86.

[37] Wabinga HR. *Helicobacter pylori* and histopathological changes of gastric mucosa in Uganda population with varying prevalence of stomach cancer. Afr Health Sci 2005;5:234–7.

[38] Tegtmeyer N, Wessler S, Backert S. Role of the cag-pathogenicity island encoded type IV secretion system in *Helicobacter pylori* pathogenesis. FEBS J 2011;278:1190–202.

[39] Yamaoka Y. Mechanisms of disease: *Helicobacter pylori* virulence factors. Nat Rev Gastroenterol Hepatol 2010;7:629–41.
[40] Kidd M, Lastovica A, Atherton J, et al. Conservation of the *cag* pathogenicity island is associated with *vacA* alleles and gastroduodenal disease in South African *Helicobacter pylori* isolates. Gut 2001;49:11–7.
[41] van Doorn LJ, Figueiredo C, Sanna R, Plaisier A, Schneeberger P, de Boer W, Quint W. Clinical relevance of the *cag*A, *vac*A, and *ice*A status of *Helicobacter pylori*. Gastroenterology 1998;115:58–66.
[42] Yamaoka Y, Kodama T, Gutierrez O, Kim JG, Kashima K, Graham DY. Relationship between *Helicobacter pylori ice*A, *cag*A, and *vac*A status and clinical outcome: studies in four different countries. J Clin Microbiol 1999;37:2274–9.
[43] Sugimoto M, Yamaoka Y. The association of *vac*A genotype and *Helicobacter pylori*-related disease in Latin American and African populations. Clin Microbiol Infect 2009;15:835–42.
[44] Louw JA, Kidd MS, Kummer AF, et al. The relationship between *Helicobacter pylori* infection, the virulence genotypes of the infecting strain and gastric cancer in the African setting. Helicobacter 2001;6:268–73.
[45] Secka O, Antonio M, Berg DE, et al. Mixed infection with *cagA* positive and *cagA* negative strains of *Helicobacter pylori* lowers disease burden in the Gambia. PLoS One 2011;6:e27954.
[46] Tanih NF, McMillan M, Naidoo N, et al. Prevalence of *Helicobacter pylori vacA, cagA* and *iceA* genotypes in South African patients with upper gastrointestinal diseases. Acta Trop 2010;116:68–73.
[47] Mistry R, Berg D, Mukhopayay AR, Segal I. *Helicobacter pylori* (Hp) virulence factors vacuolating cytotoxin gene (Vac A) genotypes and cytotoxin associated gene A (Cag A) in adult Sowetans. S Afr Med J 1999;89:888.
[48] idd M, Peek R, Lastovica A, et al. Analysis of *iceA* genotypes in South African *Helicobacter pylori* strains and relationship to clinically significant disease. Gut 2001;49:629–35.
[49] Malfertheiner P, Megraud F, O'Morain CA, et al. Management of *Helicobacter pylori* infection–the Maastricht IV Florence consensus report. Gut 2012;61:646–64.
[50] Aboderin OA, Abdu AR, Odetoyin BW, et al. Antibiotic resistance of *Helicobacter pylori* from patients in Ile-Ife, South-west, Nigeria. Afr Health Sci 2007;7:143–7.
[51] Camargo MC, Garcia A, Riquelme A, et al. The problem of *Helicobacter pylori* resistance to antibiotics: a systematic review in Latin America. Am J Gastroenterol 2014;109:485–95.
[52] Samie A, Tanih NF, Ndip RN. *Helicobacter pylori* infection-challenges of antimicrobial chemotherapy and emergence of alternative treatments. In: Trends in *Helicobacter pylori* infection. Rijeka, Croatia: IntechOpen; 2014.
[53] Onyekwere CA, Odiagah JN, Igetei R, Duro Emanuel AO, Ekere F, Smith S. Rabeprazole, clarithromycin, and amoxicillin *Helicobacter pylori* eradication therapy: report of an efficacy study. World J Gastroenterol 2014;20:3615–9.
[54] Li L, Li L, Zhou X, et al. *Helicobacter pylori* infection is associated with an increased risk of hyperemesis gravidarum: a meta-analysis. Gastroenterol Res Pract 2015;2015:278905. https://doi.org/10.1155/2015/278905.
[55] Malande OO. Symptomatic infantile *Helicobacter pylori* gastritis infection in indigenous african infants: a case series. Pan Afr Med J 2014;19:83.
[56] Pelser HH, Househam KC, Joubert G, et al. Prevalence of *Helicobacter pylori* antibodies in children in Bloemfontein, South Africa. J Pediatr Gastroenterol Nutr 1997;24:135–9.
[57] Ally R, Morton D, Bardhan K, et al. *Helicobacter pylori* (HP) in Soweto, South Africa: (I) prevalence in asymptomatic children (II) prevalence of *cag*A status in adults undergoing endoscopy. Gastroenterologia 1998;114:A55.
[58] Samie A, Obi C, Barrett L, et al. Prevalence of *Campylobacter* species, *Helicobacter pylori* and *Arcobacter* species in stool samples from the Venda region, Limpopo, South Africa: studies using molecular diagnostic methods. J Inf Secur 2007;54:558–66.

[59] Dube C, Nkosi T, Clarke A, et al. *Helicobacter pylori* antigenemia in an asymptomatic population of Eastern Cape Province, South Africa: public health implications. Rev Environ Health 2009;24:249–55.

[60] Ndip RN, Malange Takang AE, Akoachere J-FT, et al. *Helicobacter pylori* antigens in the faeces of asymptomatic children in the Buea and Limbe health districts of Cameroon: a pilot study. Tropical Med Int Health 2004;9:1036–40.

[61] Oluwasola A, Ola S, Saliu L, et al. *Helicobacter pylori* infection in South Nigerians: a serological study of dyspeptic patients and healthy individuals. West Afr J Med 2001;21:138–41.

[62] Langat AC, Ogutu E, Kamenwa R, et al. Prevalence of *Helicobacter pylori* in children less than three years of age in health facilities in Nairobi Province. East Afr Med J 2006;83(9):471–7.

[63] Ndip RN, Malange Takang AE, Ojongokpoko JE, et al. *Helicobacter pylori* isolates recovered from gastric biopsies of patients with gastro-duodenal pathologies in Cameroon: current status of antibiogram. Tropical Med Int Health 2008;13:848–54.

[64] Baako B, Darko R. Incidence of *Helicobacter pylori* infection in Ghanaian patients with dyspeptic symptoms referred for upper gastrointestinal endoscopy. West Afr J Med 1996;15:223–7.

[65] Afihene MK, Denyer M, Amuasi JH, et al. Prevalence of *Helicobacter pylori* and endoscopic findings among dyspeptics in Kumasi, Ghana. Open Sci J Clin Med 2014;2:63–8.

[66] Abebaw W, Kibret M, Abera B. Prevalence and risk factors of *H. pylori* from dyspeptic patients in Northwest Ethiopia: a hospital based cross-sectional study. Asian Pac J Cancer Prev 2014;15:4459–63.

[67] Kimang'a AN, Revathi G, Kariuki S, et al. *Helicobacter pylori*: prevalence and antibiotic susceptibility among Kenyans. S Afr Med J 2010;100:53–7.

[68] Kimang'a AN. Prevalence of *Helicobacter pylori cagA* and *vacA* allelic variants and associated disease outcomes in Kenyan patients with dyspepsia. S Afr J Epidemiol Infect 2013;28:112–6.

[69] Tadege T, Mengistu Y, Desta K, et al. Seroprevalence of *Helicobacter pylori* infection and its relationship with ABO blood groups. Ethiop J Heal Dev 2005;19:55–9.

[70] Oling M, Odongo J, Kituuka O, et al. Prevalence of *Helicobacter pylori* in dyspeptic patients at a tertiary hospital in a low resource setting. BMC Res Notes 2015;8:256–61.

[71] Tanih NF, Okeleye BI, Ndip LM, et al. *Helicobacter pylori* prevalence in dyspeptic patients in the Eastern Cape province: race and disease status. S Afr Med J 2010;100:734–7.

[72] Ayana SM, Swai B, Maro V, et al. Upper gastrointestinal endoscopic findings and prevalence of *Helicobacter pylori* infection among adult patients with dyspepsia in Northern Tanzania. Tanzan J Health Res 2014;16:16–22.

[73] Arun V, Mchembe M. Endoscopic findings and prevalence of *Helicobacter pylori* in dyspeptic patients referred for upper gastrointestinal endoscopy in Dar-Ee-Salaam. PMJUMU 2012;2(1).

[74] Asrat D, Nilsson I, Mengistu Y, et al. Prevalence of *Helicobacter pylori* infection among adult dyspeptic patients in Ethiopia. Ann Trop Med Parasitol 2004;98:181–9.

[75] Kibru D, Gelaw B, Alemu A, et al. *Helicobacter pylori* infection and its association with anemia among adult dyspeptic patients attending Butajira Hospital, Ethiopia. BMC Infect Dis 2014;14:656–62.

[76] Taddesse G, Habteselassie A, Desta K, et al. Association of dyspepsia symptoms and *Helicobacter pylori* infections in private higher clinic, Addis Ababa. Ethiop Med J 2011;49:109–16.

[77] Taddesse E, Daka D, Yemaneh D, et al. Seroprevalence of *Helicobacter pylori* infection and its related risk factors in symptomatic patients in southern Ethiopia. BMC Res Notes 2014;7:834–8.

[78] Ndububa D, Agbakwuru A, Adebayo R, et al. Upper gastrointestinal findings and incidence of *Helicobacter pylori* infection among Nigerian patients with dyspepsia. West Afr J Med 2001;20:140–5.

[79] Olokoba A, Gashau W, Bwala S, et al. *Helicabacter pylori* infection in Nigerians with dyspepsia. Ghana Med J 2013;47:79–81.

[80] Solomon O, Ajayi A. Risk factors for un-investigated dyspepsia among primary care patients in Northern Nigeria. Afr Health Sci 2013;13:1007–11.

[81] Otegbayo J, Oluwasola O, Yakubu A, et al. *Helicobacter pylori* serology and evaluation of gastroduodenal disease in Nigerians with dyspepsia. Afr J Clin Exp Microbiol 2004;5:131–8.
[82] Archampong TNA, Asmah RH, Wiredu EK, et al. Epidemiology of *Helicobacter pylori* infection in dyspeptic Ghanaian patients. Pan Afr Med J 2015;20:178.
[83] Gangaidzo I, Mason P, Kiire C, et al. *Helicobacter pylori* in endoscopy patients in Zimbabwe: value of enzyme-linked immunosorbent assay and a rapid urease test. Trans R Soc Trop Med Hyg 1995;89:502–5.
[84] Walker TD, Karemera M, Ngabonziza F, et al. *Helicobacter pylori* status and associated gastroscopic diagnoses in a tertiary hospital endoscopy population in Rwanda. Trans R Soc Trop Med Hyg 2014;108:305–7.
[85] Mathewos B, Moges B, Dagnew M. Seroprevalence and trend of *Helicobacter pylori* infection in Gondar University Hospital among dyspeptic patients, Gondar, North West Ethiopia. BMC Res Notes 2013;6:346.
[86] Mustapha S, Ajayi N, Nggada H, et al. Endoscopic findings and the frequency of *Helicobacter pylori* among dyspeptic patients in Maiduguri, North-Eastern Nigeria. HMRJ 2008;5:78–81.
[87] Ogutu EO, Kang'ethe SK, Nyabola L, Fau-Nyong'o A, et al. Endoscopic findings and prevalence of *Helicobacter pylori* in Kenyan patients with dyspepsia. East Afr Med J 1998;72:85–9.
[88] Wafula J, Lule G, Otieno C, et al. Upper gastrointestinal findings in diabetic outpatients at Kenyatta National Hospital, Nairobi. East Afr Med J 2002;79:232–6.
[89] Segamwenge IL, Kagimu M, Ocama P, et al. The utility of the *Helicobacter pylori* stool antigen test in managing dyspepsia: an experience from a low resource setting. Afr Health Sci 2014;14:829–34.
[90] Ugiagbe R, Omuemu C. Non-ulcer dyspepsia: an endoscopic review. Afr J Med Health Sci 2013;12:6.
[91] Jemilohun AC, Otegbayo JA, Ola SO, et al. Prevalence of *Helicobacter pylori* among Nigerian patients with dyspepsia in Ibadan. Pan Afr Med J 2010;6:6–9.
[92] Muhumba E, Magafu MG, Mwita JC, et al. Gastrointestinal endoscopy at Amana municipal referral Hospital in Tanzania: reasons for referral and findings. Tanzan J Health Res 2015;2:17. https://doi.org/10.4314/thrb.v17i2.10.
[93] Mothes H, Chagaluka G, Chiwewe D, et al. Do patients in rural Malawi benefit from upper gastrointestinal endoscopy? Trop Dr 2009;39:73–6.
[94] Palmer D, Watson K, Allen M. *Helicobacter pylori* infection and peptic ulcer disease in Cameroon, West Africa. J Clin Gastroenterol 1994;18:162–3.
[95] O'Keefe S, Salvador B, Nainkin J, et al. Empiric treatment based on *Helicobacter pylori* serology cannot substitute for early endoscopy in the management of dyspeptic rural black Africans. S Afr Med J 2000;90:1129–35.
[96] Abdalazeem N, Abdul-Aziz H, Adam AA, et al. Clarithromycin-resistant *Helicobacter pylori* strains among dyspeptic patients in Sudan. Al-Neelain Medical Research Centre; 2012.
[97] Gyedu A, Yorke J. Upper gastrointestinal endoscopy in the patient population of Kumasi, Ghana: indications and findings. Pan Afr Med J 2014;18:327.
[98] Asrat D, Nilsson I, Mengistu Y, et al. Prevalence of *Helicobacter pylori* vacA and cagA genotypes in Ethiopian dyspeptic patients. J Clin Microbiol 2004;42:2682–4.
[99] Owen RJ, Xerry J, Gotada, et al. Analysis of geospecific markers for *Helicobacter pylori* variants in patients from Japan and Nigeria by triple-locus nucleotide sequence typing. Microbiology 2004;150:151–61.
[100] Letley D, Lastovica A, Louw J, et al. Allelic diversity of the *Helicobacter pylori* vacuolating cytotoxin gene in South Africa: rarity of the *vacA* s1a genotype and natural occurrence of an s2/m1 allele. J Clin Microbiol 1999;37:1203–5.
[101] Kidd M, Lastovica A, Atherton J, et al. Heterogeneity in the *Helicobacter pylori vacA* and *cagA* genes: association with gastroduodenal disease in South Africa? Gut 1999;45:499–502.
[102] Secka O, Antonio M, Tapgun M, et al. PCR-based genotyping of *Helicobacter pylori* of Gambian children and adults directly from biopsy specimens and bacterial cultures. Gut Pathog 2011;3:5. https://doi.org/10.1186/1757-4749-3-5.

[103] Smith SI, Kirsch C, Oyedeji KS, et al. Prevalence of *Helicobacter pylori vacA*, *cagA* and *iceA* genotypes in Nigerian patients with duodenal ulcer disease. J Med Microbiol 2002;51:851–4.
[104] Breurec S, Michel R, Seck A, et al. Clinical relevance of *cagA* and *vacA* gene polymorphisms in *Helicobacter pylori* isolates from Senegalese patients. Clin Microbiol Infect 2012;18:153–9.
[105] Asrat D, Kassa E, Mengistu Y, et al. Antimicrobial susceptibility pattern of *Helicobacter pylori* strains isolated from adult dyspeptic patients in Tikur Anbassa University Hospital, Addis Ababa. Ethiop Med J 2004;42:79–85.
[106] Secka O, Berg DE, Antonio M, et al. Antimicrobial susceptibility and resistance patterns among *Helicobacter pylori* strains from the Gambia, West Africa. Antimicrob Agents Chemother 2013;57:1231–7.
[107] Seck A, Burucoa C, Dia D, et al. Primary antibiotic resistance and associated mechanisms in *Helicobacter pylori* isolates from Senegalese patients. Ann Clin Microbiol Antimicrob 2013;12:3.
[108] Seck A, Mbengue M, Gassama-Sow A, et al. Antibiotic susceptibility of *Helicobacter pylori* isolates in Dakar, Senegal. J Infect Dev Ctries 2009;3:137–40.
[109] Ani AE, Malu AO, Onah JA, Queiroz DM, Kirirschner G, Rocha GA. Antimicrobial susceptibility test of *Helicobacter pylori* isolated from Jos, Nigeria. Trans R Soc Trop Med Hyg 1999;93:659–661S.
[110] Smith S, Oyedeji KS, Arigbabu AO, et al. High amoxycillin resistance in *Helicobacter pylori* isolated from gastritis and peptic ulcer patients in western Nigeria. J Gastroenterol 2001;36:67–8.
[111] Tanih NF, Okeleye BI, Naidoo N, et al. Marked susceptibility of South African *Helicobacter pylori* strains to ciprofloxacin and amoxicillin: clinical implications. S Afr Med J 2010;100:49–52.
[112] Tanih NF, Ndip RN. Molecular detection of antibiotic resistance in South African isolates of *Helicobacter pylori*. Gastroenterol Res Pract 2013;2013(2013):259457. https://doi.org/10.1155/2013/259457.

FURTHER READING

[113] Louw JA, Jaskiewicz K, Girdwood A, et al. *Helicobacter pylori* prevalence in non-ulcer dyspepsia-ethnic and socio-economic differences. S Afr Med J 1993;83:169–71.

CHAPTER

PANCREATITIS IN SUB-SAHARAN AFRICA

10

Isidor Segal

G.I.T. Unit, Chris Hani Baragwanath Hospital and University of Witwatersrand, Soweto, South Africa

INTRODUCTION

Economic development and countries growth from low-income to middle-income status will probably lead to decreases in poverty rates in sub-Saharan Africa (SSA). Economic growth however is accompanied by rapid urbanization, increased alcohol and tobacco consumption, and adverse changes in dietary patterns. These changes in lifestyle and habitation will have consequences for human health and in particular to the increase in pancreatitis. A model for these changes is illustrated by what has occurred in Soweto, South Africa. Soweto is served by the Chris Hani Baragwanath Hospital. This hospital with 3200 beds is one of the largest in the world [1]. The current population of Soweto is approximately four million, a complex polyglot of largely immigrant and mainly lower working class people but with an emerging highly urbanized and better trained upper working class and middle class [2].

Prior to 1962, Black people in South Africa were legally forbidden to purchase Western-type alcohol. Following repeal of this legislation, a markedly increased consumption of spirits and fortified wine occurred with a corresponding decrease in traditional beverages [3,4]. Twenty years after the repeal of this legislation, pancreatitis emerged and has become common in Black people of Soweto [1]. Increased admissions for pancreatitis occurred some 20 years after Africans had access to Western-type alcohol. Case-control studies showed that approximately two-thirds still drank mainly home-consumed traditional home-brewed beer of a low alcohol content.

25 patients with calcific chronic pancreatitis (CCP) were diagnosed in 1977. In the 3-year period 1981–83, there were 55 new cases. A further survey in 1988 identified 90 new cases. Male-to-female ratio was 6:1.The mean age of the patients was 40 years and the mean alcohol intake was 150 g/day for an average of 15 years. In 1994, 136 patients with acute pancreatitis (AP) were admitted to Baragwanath Hospital, Soweto. Alcohol was the predominant etiological factor in 83% patients and biliary disease in 7%. There was a high frequency of pulmonary tuberculosis and diabetes diagnosed at initial presentation. The clinical features of patients with pancreatitis illustrated the ferocity of the disease. Mortality was high—15%. There is also a lingering morbidity in a substantial number of patients. The morbidity and mortality is an underestimate as many of the patients are lost to follow up [1].

It is that the small population of drinkers in Africa have excessively heavy drinking habits, episodes often found in youth drinking [5,6].

Of recent interest is the striking shift in disease pattern with regard to biliary pancreatitis (BP). Although alcohol is still the major etiological factor, choledocholithiasis has become more common than before. This is in accord with urbanization and dietary changes that have occurred in SSA. As

Digestive Diseases in Sub-Saharan Africa. https://doi.org/10.1016/B978-0-12-815677-3.00010-4

a result, laparoscopic cholecystectomy has become a common operation at Baragwanath Hospital. Another reason for the increase are the number of patients with HIV treated using the HAART regime. The result is an increased number of patients with BP (Personal Communication-Prof. M. Smith).

CLINICAL ASPECTS [1,7]

AP manifests as an episode of severe abdominal pain with hyperamylassemia. The mild edematous (interstitial inflammation) form is a self-limiting disease with <1% mortality, whereas the necrotizing form with local necrotizing inflammation, systemic, and local complications, which may be sterile or infected. Infected AP carries a high mortality—up to 20%. In over 95% of cases, the characteristic abdominal pain is the presenting feature. The typical pancreatic pain is epigastric radiating to the back, associated with nausea and vomiting, aggravated by food intake. The discomfort is lessened by sitting up and leaning forward and by fasting.

On clinical examination there may be mild-to-severe tenderness in the epigastrium. Patients with more severe forms of pancreatitis are acutely ill. Low-grade fever, hypotension, and tachycardia may be present. Initial examination of abdomen typically reveals distention and hypoactive or absent bowel sounds. Clinical evidence for pleural effusion, atelectasis, pneumonia, scleral icterus may be seen. Gray Turner sign (ecchymotic discoloration of the flanks) or Cullen sign (periumbilical) are extremely rare and are not needed for diagnosis, but are among the list of single markers of severity.

DIAGNOSIS

The differential diagnosis includes cholangitis, cholecystitis, peptic ulcer perforated peptic ulcer, and myocardial infarction.

Initial diagnostic tests

In AP, the levels of serum amylase increase sharply after the onset of symptoms and return to normal within 4 days. Elevation of amylase level > 3 times normal is diagnostic.

Elevation of serum lipase levels is often considered to be more specific. However, it also occurs in other conditions. Lipase levels rise within 8 h of onset, peak in 24 h, and decrease within eight to 14 days.

Once a diagnosis is established, the next step is to predict the severity (risk stratification). A number of well-known criteria based on multiple scoring—Ranson, Imrie, Glasgow, APACHE II scale, and other single markers of prognosis are available.

IMAGING STUDIES

An initial plain X-ray of the abdomen helps more to exclude the diagnosis of perforated ulcer or intestinal obstruction rather than to diagnose pancreatitis. Dilated loops of small bowel ("sentinel loop") or transverse colon ("colon-cut off sign") are nonspecific and infrequent. Chest X-ray helps to evaluate the lungs and to diagnose pleural effusion. An initial abdominal ultrasound is a must in all cases except when an alcoholic etiology is quite clear. The main purpose is to look for gallstones or sludge, and assess the CBD size. Contrast enhanced CT helps in assessing severity of AP.

Specialized studies such as magnetic resonance imaging (MRI), magnetic resonance cholangiopancreatography (MRCP), and endoscopic ultrasound (EUS) have only limited value in the routine management, but may be needed to exclude a CBD stone in the presence of a dilated duct or persistently abnormal liver enzymes.

MANAGEMENT OF ACUTE PANCREATITIS

The principles of management include (1) giving "rest to the pancreas"; (2) management of pain; (3) replacement of fluid and electrolytes; (4) adequate nutritional support; (5) early recognition of severity; and (6) identification of etiology and definitive management of the precipitating factor (i.e., gallstones) and prompt recognition of complications that need specialized management.

Prognosis

Mortality in AP occurs in two peaks. The first peak is usually in the first 72 h and in a small amount of cases occurs even before the patient reaches the hospital. The first peak of mortality cannot be assessed by imaging studies since it is not morphological, but due to cytokine activated and systemic inflammatory response, and multiorgan failure. Patients suffering from early and persistent multiorgan failure syndrome (organ failure that has persisted for >48 h or more than two systems failed) have a high risk of mortality (over 30%).

The second peak that is associated with necrosis (irreversible death of tissues) and/or infection occurs 2 weeks later. Gross destruction of the pancreas (necrosis) is seen in about 20% of patients. The pancreas fails to enhance after a rapid intravenous bolus injection of contrast followed by rapid CT imaging. Patients with sterile necrosis also display clinical signs resembling sepsis characterized by fever, leukocytosis, and abdominal pain. Patients with severe pancreatic necrosis (>50%) are at high risk for infection. A CT-guided fine needle aspiration (FNA) and gram staining of the aspirated material helps in the diagnosis of infected necrosis. Infected necrosis is treated with antibiotics and by surgical debridement. Ciprofloxacin with metronidazole or imipenem (an intravenous B-lactam antibiotic) are among the choice of antibiotics.

Complications

AP is a multisystem disease. The complications can be classified as intra-abdominal and systemic.

Fluid collections occur in both edematous and necrotizing pancreatitis. Most of the pancreatic fluid collections do not need interventions.

A pseudocyst that usually forms after 4 weeks may be asymptomatic or symptomatic. Asymptomatic pseudocyst needs only observation. A pseudocyst that is painful, infected, pressing upon adjacent organs or growing under observation requires surgery. Pancreatic abscess needs surgery and antibiotic therapy. An open surgical debridement with continuous short-term lavage of the lesser sac is needed. Repeated surgery may be needed. The prognosis is better than infected necrosis.

WHEN TO REFER THE PATIENT?

A majority of patients have mild (edematous) pancreatitis. Patients with markers of severity, prolonged biliary obstruction, signs and symptoms of sepsis, imaging evidence of necrosis, need to be referred to a gastroenterologist, surgeon as well as intensive care specialist. Severe AP is to be managed by a team of physicians consisting of a gastroenterologist, surgeon, an internist, an infectious disease specialist, and interventional radiologist. Patients with severe AP need intensive care management.

CONCLUSIONS

The commonest causes of AP are gallstones (biliary sludge, microlithiasis) and chronic alcoholism. The typical history is of epigastric pain often radiating to the back and elevation of serum amylase and lipase 3 times above normal levels. Imaging procedures include abdominal ultrasound (always) and abdominal contrast enhanced CT Scan usually after 72h. Severity is predicted by various scoring systems including Ransons, APACHE 2, and Glasgow score. In mild cases, management focuses on NPO,IV fluids, and analgesia. Severe cases require ICU care as multiple organ dysfunction can occur. The latter complication can also occur after the second week following the onset. In those patients requiring prolonged nutritional support, nasojejunal feeding is the preferred method. Surgery is required for BP and in infected necrosis.

A retrospective diagnosis of AP is properly retained only when morphological and functional tests of the exocrine pancreas are abnormal.

CHRONIC PANCREATITIS

Chronic pancreatitis (CP) is characterized clinically by recurrent upper abdominal pain, associated with malabsorption of fat(steatorrhea) and diabetes mellitus [8].

ETIOLOGY (BRAGANZA-MANCHESTER HYPOTHESIS) [9]

Case-control studies in AP and CP identified the same three environmental factors in each disease, namely, heavy alcohol consumption, marked exposure to occupational chemicals, and low intake of fruit (a major source of Vit.C). This suggests that the two diseases may be part of a pathobiological spectrum that is linked by pancreatic oxidant stress [1]. Further, asymptomatic chronic alcoholics had plasma glutathione concentrations that were midway between the values in nonalcoholic controls and patients with CP. And finally apparently healthy Sowetans were in a state of oxidant stress that was tied in with their very poor vitamin C status and lower selenium status than in the U.K. These data, and evidence that both antioxidants mitigate against alcoholic toxicity in experimental studies, may offer scope for disease prophylaxis in this underprivileged community [2,9,10].

Braganza [2,10] has proposed a hypothesis that Cytochrome P450 (CYP) induction in the liver and the entry of toxic xenobiotic metabolites intermediates into the pancreas by way of refluxed bile or the bloodstream initiated CP irrespective of putative etiological factor, and also nongallstone AP. It has been shown that the toxicity of ethanol, especially when combined with high dietary unsaturated fat, is related to the production of free radical intermediates and that this combination now provides a reproducible animal model of CP.

HIV/AIDS AND PANCREATITIS

UNAIDS estimates that in 2014 a total of 34.2 million people were living with HIV infection.

SSA is the most affected continent. In Southern Africa HIV is hyperendemic, with an overall prevalence among adults of up to 31% in Swaziland, 25% in Botswana, and 17% in South Africa [11].

HIV disease can present initially as pancreatitis [12]. The disease can also result from therapy used to treat HIV disease [13]. The clinical presentation of pancreatitis in HIV-positive patients is no different to that of the general population except that other manifestations of HIV may be present [13]. In addition, comorbid conditions such as alcohol, biliary disease, and opportunistic infections may increase the prevalence of pancreatitis in this cohort [14].

The initiation of ARP therapy and HAART (highly active antiretroviral therapy) medications that may cause pancreatitis are [14]: Didanosine, Stavudin, the protease inhibitor, Lopinovir/titanovir. In children. Lamivudine may also cause pancreatitis.

BILIARY PANCREATITIS

Biliary lithiasis is the commonest cause of AP [15] and has been increasing over the past two decades. Gallstones account for at least half of the 4.8–24.2 cases of pancreatitis per 100,000 people that occur in Western countries. Approximately 80,000 cases occur in the USA; 17 per 100,000 new cases. In Japan, the annual incidence lies between 5 and 80 per 100,000 of the population [16].

AP caused by gallstones was considered rare in SSA with alcohol as the main etiologic factor. The lowest rates for gallstone disease were reported to be in African populations. However, sickle cell disease may result in pigment gallstones and prophylactic cholecystectomy is recommended in this group [17]. It is noteworthy that gallstones are becoming more frequent in the urban Black population of South Africa and as a result BP has become more common. Laparoscopic cholecystectomy is probably one of the more common operations carried out at Baragwanath Hospital (Personal communication. Prof. M. Smith).

CLINICAL SIGNS AND SYMPTOMS

AP is characterized clinically by an ill patient with sudden onset of severe abdominal pain and severe vomiting. Generalized tenderness of the abdomen is present but abdominal rigidity unusual. Jaundice may be present. Patient should be examined specifically for CVS, respiratory and renal dysfunction.

Document any presence of organ failure [18]. At this stage, it is reasonable to do the Apache 11 Score.

Laboratory investigations

Serum amylase increases in 3–10 days. Amylase lacks specificity and can be raised in other disorders [16].

Serum lipase is more specific and has a longer half-life but can be raised in perforated peptic ulcer, mesenteric vascular occlusion, and in intestinal obstruction2.

Review serum bilirubin before confirming diagnosis.

Liver function tests. AST, ALT, Gamma GT.

Fasting serum calcium and lipid profile.

CRP—indicates severity of pancreatitis >48 h after onset of symptoms.

BMI and standard blood tests.

ASSESSMENT OF SEVERITY USING PROGNOSTIC SCORING SYSTEMS

Glasgow (Imrie) system better for gallstone pancreatitis than Ransom criteria [16].

Apache 11 Score

Ransom criteria

Radiology investigations

X-ray chest

X-ray abdomen

Ultrasound of the abdomen. Note CT often less sensitive than ultrasound [16].

EUS—more sensitive than either abdominal US or CT to detect common duct stones.

CT scan allows for identification of pancreatic edema, fluid, or cysts and the severity of pancreatitis for grading purposes. CT also detects complications, including development of pseudocysts abscess, necrosis hemorrhage, and vascular occlusion [16].

Urgent therapeutic ERCP should be performed in gall stone pancreatitis of suspected or proven gallstone etiology [17] who satisfy the criteria for predicted or actual severe pancreatitis; or when there is cholangitis,early jaundice, or a dilated common bile duct [17]. Procedure is best carried out in the first 72 h after onset of pain. All patients undergoing early ERCP for severe gall stone pancreatitis require endoscopic sphincterotomy whether or not stones are found in the common bile duct [19].

Patients with signs of cholangitis require endoscopic sphincterotomy or duct drainage by stenting to ensure relief of biliary obstruction.

MRCP does not have the complications of ERCP but is not reliable if stones are <4 mm. Also extraction of stones is not possible. MRCP can assess pancreatic and peripancreatic cysts [16].

SCORING SYSTEMS—PREDICT SEVERITY

Glasgow (Imrie) system is better for gallstone pancreatitis than Ransom criteria.

Apache score.

MANAGEMENT

AP in most patients is mild to moderate.

Fluid rehydration, pain control, and provision of oxygen are the initial treatment [18]. The patient should be monitored for any signs of early organ failure such as hypotension, pulmonary or renal insufficiency. At this stage, a plain X-Ray and ultrasound of the abdomen should be carried out. If there are any signs of early organ failure, transfer to the ICU is advisable.

In ICU, a central line should be inserted and an MRI and CT scan arranged. If a common bile duct stone is observed, urgent ERCP, shincterotomy, and removal of stone should be carried out.

It is important to have a team of gastroenterologist/surgeon caring for the patient.

Necrosectomy is indicated for infected or sterile necrosis; if there are persistent clinically significant symptoms, enteral nutrition should be instituted [20].

A laparoscopic cholecystectomy should be performed on all patients with mild disease prior to hospital discharge as they have a 30%–40% risk of recurrent attacks [20]. Alexakis and Neoptolomos [20] advise routine cholangiography during laparoscopic cholecystectomy followed by selective postoperative endoscopic sphincterotomy. Laparoscopic common bile duct exploration and postoperative ERCP are safe and reliable in clearing common bile duct stones. They further advise that surgically fit patients who had undergone endoscopic sphincterotomy stone extraction and have concomitant cholelithiasis should undergo cholecystectomy.

Twenty to thirty percent of patients have severe disease. The course is biphasic. The first phase (1–2 weeks) is characterized by early toxic enzymatic action and evolution in SIRS-MODS-{Multi Organ Dysfunction Syndrome}. The latter phase (3–4 weeks) [15] is characterized by septic complications of pancreatitis and results in pancreatic-peripancreatic fluid necrotic collections [15].

Long-term follow up is necessary to monitor patients for any complications that may occur.

MAJOR COMPLICATIONS [21]

Necrosis (sterile or infected)
Pseudocyst
Abscess
GIT bleeding
Splenic injury
Organ dysfunction

CONCLUSIONS

Biliary lithiasis is the commonest cause of BP.

Patient should examined specifically for CVS, respiratory and renal dysfunction.

Raised amylase levels are a good index for the diagnosis of AP, but serum lipase is more specific and remains raised for a longer period of time.

The Glasgow (Imrie) scoring system is better for gallstone pancreatitis than Ransom's criteria.

Abdominal ultrasound is more sensitive than CT scan, but EUS is better diagnostically than abdominal ultrasound or CT scan.

Urgent therapeutic ERCP should be performed in gall stone pancreatitis of suspected or proven gallstone etiology. The procedure is best carried out in the first 72h after onset of pain. All patients undergoing early ERCP for severe gallstone pancreatitis require endoscopic sphincterotomy whether or not stones are found in the common bile duct. Patients with signs of cholangitis require endoscopic sphincterotomy or duct drainage by stenting to ensure relief of biliary obstruction.

A laparoscopic cholecystectomy should be performed in all patients with mild disease prior to hospital discharge.

Follow up is important in order to detect and treat any complications that may arise.

TROPICAL PANCREATITIS

Tropical pancreatitis is a form of CP characterized by young age of onset, recurrent abdominal pain, malnutrition, rapid progression, multiple, large, pancreatic calculi, and diabetes mellitus, occurring mostly among children and young adults of many Afro-Asian nations. There is an absence of a history of alcoholism or biliary tract disease [22,23].

EPIDEMIOLOGY

The syndrome of CP with pancreatic calculi and diabetes has been reported by different observers from many countries such as Uganda, Nigeria, the Republic of Congo, Malawi, Zambia, Ghana, Ivory Coast, and Madagascar on the African continent [22].

PATHOLOGY

The histologic changes in the pancreas are similar to those of alcoholic pancreatitis. In advanced stages of the disease, the pancreatic gland is small and the surface irregular and nodular because of fibrosis. Microscopically, the characteristic feature is diffuse fibrosis of the pancreas. Pancreatic calculi vary in color, size, and shape. As the disease advances, the islets become atrophic and are isolated and surrounded by dense fibrous tissue [22]. There is a high prevalence of pancreatic carcinoma suggesting that tropical pancreatitis is a premalignant disease.

ETIOLOGY AND PATHOGENESIS

The exact etiology of this disease has not been established. CP, alcoholic or tropical, has been hypothesized as one of the many diseases caused by unmitigated free radical injury. An association with a high consumption of diet rich in cassava root (tapioca, *Manihot esculenta*) and tropical pancreatitis was suggested in many previous reports. However, the association appears to be weak. Many recent studies have looked for genetic abnormalities in all forms of CP following the discovery of genetic mutations in hereditary pancreatitis. In the normal pancreas, a number of mechanisms work synergistically preventing the premature activation of trypsinogen to trypsin. In four studies, three from India and the other from Bangladesh, tropical pancreatitis was noted to be associated with the SPINK 1 N34S mutation [22].

CLINICAL PRESENTATION [22,23]

Malnutrition may be present. This is either primary malnutrition or secondary to CP. Diabetes is a common presentation. Improvement in the management of diabetes has resulted in a longer life span not noted in earlier observations. Patients initially may seek medical attention for diabetes mellitus, which becomes clinically manifest a few years after the onset of pancreatalgia. A pain-free period of 1–2 years and an apparent transient improvement in the clinical picture prior to the onset of diabetes are not unusual. Overt exocrine pancreatic insufficiency characterized by steatorrhea is the least striking clinical feature, attributable to the very low consumption fat in the diet.

Pancreatic carcinoma is a common complication.

RADIOLOGY

The disease is characterized by pancreatic calculi on plain X-Ray of the abdomen.

TREATMENT

The main object of treatment is control of the underlying diabetes.

REFERENCES

[1] Segal I. Pancreatitis in Soweto, South Africa. Focus on alcohol-related disease. Digestion 1998;59(Suppl 4):25–35.

[2] Segal I, Dubb AA, Ou Tim L, et al. Duodenal ulcer and working class mobility in an African population in South Africa. Br Med J 1978;(i)469–71.

[3] Bradshaw E, Schonfeld M. Smoking, drinking and oesophageal cancer in African males of Johannesburg, South Africa. Br J Cancer 1974;30:157.

[4] Van Rensburg SJ. Epidemiology and dietary evidence for specific nutritional predisposition to oesophageal cancer. J Natl Cancer Inst 1981;67:243.

[5] Seggie J. Alcohol and South Africa's youth. S Afr Med J 2012;102(7).

[6] Obot IS. Alcohol use and related problems in sub-Saharan Africa. Afr J Drug Alcohol Stud 2006;5(1):17–26.

[7] Pitchumoni CS. Acute pancreatitis. In: Segal I, Pitchumoni CS, Sung J, editors. Gastroenterology and hepatology manual. vol. 29. Australia: McGraw-Hill; 2011. p. 304–27.

[8] Pitchumoni CS. Chronic pancreatitis. In: Segal I, Pitchumoni CS, Sung J, editors. Gastroenterology and hepatology manual. vol. 29. Australia: McGraw-Hill; 2011. p. 312–27.

[9] Braganza JM. The pancreas. Recent Adv Gastroenterol 1996;6:251–80.

[10] Braganza JM, Schofield D, Snehalatha, Mohan V. Micronutrient antioxidant status in tropical compared with temperate-zone chronic pancreatitis. Scand J Gastroenterol 1993;2:1098.

[11] Piot P, Quinn TC. Response to AIDS pandemic—a global health model. N Engl J Med 2013;368:2210–8.

[12] Gazanfar A, Kumar M, Dhar S, et al. HIV and pancreatitis. JK Sci 2001;3(4):188–90.

[13] Dragovic G. Acute pancreatitis in HIV/AIDS patients: an issue of concern. Asian Pac J Trop Med 2013;3(6):422–5.

[14] Blake M, Sherer R. Management of the adverse effects of antiretroviral therapy and medication adherence. Clin Infect Dis 2000;30:S96–116.

[15] Vincenzo N. Management of acute biliary pancreatitis. In: Copeland S, editor. Acute pancreatitis: Health effects, clinical aspects and emerging therapies. Nova Science Publisher Inc.; 2016.

[16] Hazem ZM. Acute biliary pancreatitis: diagnosis and treatment. Saudi J Gastroenterol 2009;15(3):147–55.

[17] Njeze GE. Gallstones. Niger J Surg 2013;19(2):49–55.

[18] Weinstein WM, Hawkey CJ, Bosch J. Chapter 71. Elsevier Mosby; 2005505–12.

[19] Working Party of the British Society of Gastroenterology, Association of Surgeons of Great Britain and Ireland, Pancreatic Society of Great Britain and Ireland, Association of Upper GI Surgeons of Great Britain and Ireland. UK guidelines for the management of acute pancreatitis. Gut 2005;54:iii1–9. https://doi.org/10.1136/gut.2004.057026.

[20] Alexakis N, Neoptolomos JP. Algorithm for the diagnosis and treatment of acute biliary pancreatitis. Scand J Surg 2005;94:124–9.

[21] Steinberg WM. Acute pancreatitis. In: Sleisenger and Fortdtran's gastrointestinal and liver disease. 8th ed. Vol 1. Philadelphia: Saunders Elsevier; 2006. p. 1241–69. [Chapter 56].

[22] Pitchumoni CS. Chronic pancreatitis. In: Segal I, Pitchumoni CS, Sung J, editors. Gastroenterology and hepatology manual. 29. Australia: McGraw-Hill; 2011. p. 312–27.

[23] Balakrishnan V, Nair P, Radhakrishnan L, Narayanan VP. Tropical pancreatitis—a distinct entity, or merely a type of chronic pancreatitis? Indian J Gastroenterol 2006;25:74–80.

CHAPTER

HIV AND THE DIGESTIVE SYSTEM

11

Alan Karstaedt

Division of Infectious Diseases, Department of Medicine, Chris Hani Baragwanath Hospital, University of the Witwatersrand, Johannesburg, South Africa

INTRODUCTION

The 4th decade of HIV infection has been the harbinger of a remarkable rise in optimism that the "end game" is approaching [1, 2]. This has been predicated on a number of factors, including the rapid expansion of the numbers of people on effective antiretroviral therapy (ART), pursuit of a cure for HIV infection, and combined prevention strategies, which promise significant reductions in rates of infection over time. There has too been the tantalizing suggestion that HIV's ability to cause AIDS may weaken over time. The success of ART has culminated in the concept of HIV infection as a chronic disease resulting in an expected lifespan, which can approach that of the non HIV-infected [3]. This depends partly on initiating ART at higher CD4 counts, on sociodemographic factors and on lifestyle and behavioral determinants including alcohol use, smoking, recreational drug use, and viral hepatitis coinfection. The success of ART has been bolstered by less toxic regimens and single tablet combination drugs improving compliance. The treatment cascade is the term being used for a conceptual model that details services across the continuum of care. This comprises testing of individuals, access to care, starting ART, staying in care, and remaining adherent to treatment [1]. Treatment does not fully restore immunity, and concerns remain over the relative increase in importance of non-AIDS conditions including chronic diseases of most organ systems including the liver and of non-AIDS defining cancers [4]. An increase in the aging population with HIV infection over time will require innovative ideas to be able to provide effective chronic care for people with the multimorbidities, which can be anticipated. New concepts in the field of prevention of infection, beyond behavioral and structural approaches and circumcision and vaginal microbiocides, include starting ART at higher CD4's or in everyone diagnosed with HIV infection, in the expectation that new infections will be significantly reduced. This is termed treatment as prevention. Preexposure prevention treatment is effective for groups such as discordant couples and those at highest risk of transmission by dint of sexual practices or injection drug use. Such putative increases in the numbers of people on treatment will demand research in implementation science and the development of models of care that can be replicated across different geographic areas. The need for expanded monitoring of efficacy of ART by viral load and CD4 count measurement and assessment of intervals for testing will impact on costs of these programs. An effective vaccine remains elusive, but promises most in the long term.

Digestive Diseases in Sub-Saharan Africa. https://doi.org/10.1016/B978-0-12-815677-3.00011-6

EPIDEMIOLOGY

Worldwide, at the end of 2013 according to UNAIDS figures, an estimated 35 million people were living with HIV infection, of whom 19 million do not know their status. These sobering figures are partially balanced by a 13% reduction in new infections, at 2.1 million worldwide, and a 35% reduction in deaths since the peak in 2005. Nearly 13 million people were accessing antiretroviral therapy (ART). Sub-Saharan Africa bears the brunt of infection, accounting for almost 71% of the total, with nearly 1 in every 21 adults living with HIV infection. 6.2 million people are infected in South Africa, which has more people with HIV than any other country. About 18% of all adults in South Africa are HIV positive. Treatment with ART has reached 37% of those in sub-Saharan Africa.

Adherence clubs have been developed as a way to reduce clinic congestion, simplify logistics, and help with motivation. The clubs encourage acceptance and openness about HIV in the community, which helps with both screening and treatment. The program offers a model for treatment in countries with limited resources [5].

ART AND THE GASTROINTESTINAL TRACT

There are a number of predictable side effects and complications of the commonly used antiretroviral drugs (Table 11.1). Many of them cause mild reversible gastrointestinal intolerance on initiation with nausea, vomiting, and diarrhea. Abacavir hypersensitivity, which can be fatal, occurs at a rate of 5%–8% at a median of 9 days after starting and usually in the first 6 weeks. Susceptibility is genetic and associated with HLA-B*5701 haplotype in the majority. Testing where available should precede abacavir use. There is marked genetic and ethnic variability of HLA-B*5701 with rates <1% in Africa. The GIT symptoms accompany fever, rash, constitutional symptoms, and possibly respiratory symptoms. All ART must be interrupted. Resolution occurs within a few days. Abacavir should be stopped and not be used for rechallenge. Lactic acidosis or symptomatic hyperlactatemia, a less severe form, is a mitochondrial toxicity caused by the nucleoside reverse transcriptase drugs stavudine, didanosine, and zidovudine and in severe cases may be accompanied by hepatic steatosis and necrotizing pancreatitis. It occurs usually 3–36 months after starting ART with one of these drugs. Lactic acidosis occurs in 1–2 patients/1000 patient-years but is more common in women. All ART drugs must be interrupted. The offending drug should be withdrawn permanently and substituted with an appropriate replacement. The hepatitis associated with the nucleoside reverse transcriptase inhibitors does not commonly result in jaundice.

Nevirapine hepatotoxicity is potentially fatal. It is a hypersensitivity reaction, which occurs in the first 6–18 weeks of treatment in 2.5%–11% of patients. Rash may coexist. Discontinue all antiretrovirals and any other potentially hepatotoxic drugs. A later more benign form of transaminitis is also encountered.

Atazanavir causes a reversible increase in indirect bilirubin that is not clinically significant and does not require discontinuation. It is caused by inhibition of UGT1A1 (UDP glucuronosyl transferase), an enzyme needed for conjugation of bilirubin, and is similar to Gilbert Syndrome. A small number of patients on lopinavir/ritonavir have persistent diarrhea, which may require drug substitution. The ritonavir-boosted protease inhibitors can cause a clinically significant hepatitis.

Table 11.1 Side Effects and Complications of Antiretroviral Therapy Involving the Gastrointestinal Tract

Drug	Side Effect
Nucleoside reverse transcriptase inhibitors	
Abacavir	Nausea, vomiting, diarrhea and abdominal pain-hypersensitivity reaction
Didanosine	Lactic acidosis ± hepatic steatosis (nausea, vomiting, abdominal pain), pancreatitis, noncirrhotic portal hypertension (nodular regenerative hyperplasia)
Emtricitabine	Risk of hepatitis B flare if stopped
Lamivudine	Risk of hepatitis B flare if stopped, pancreatitis in children
Stavudine	Lactic acidosis ± hepatic steatosis (nausea, vomiting, abdominal pain), pancreatitis, transaminitis
Tenofovir	Risk of hepatitis B flare if stopped, flatulence
Zidovudine	Lactic acidosis ± hepatic steatosis (nausea, vomiting, abdominal pain), transaminitis
Nonnucleoside reverse transcriptase inhibitors	
Nevirapine	Early hepatotoxicity; later transaminitis
Efavirenz	Transaminitis. Rarely severe hepatotoxicity
Protease inhibitors	
Atazanavir	Indirect hyperbilirubinemia
Lopinavir/ritonavir	Diarrhea, hepatitis, pancreatitis
Most PIs	Hepatitis
Chemokine coreceptor antagonists	
Maraviroc	Hepatitis

Immune reconstitution inflammatory syndrome (IRIS) occurs with restoration of the immune system usually in the first 3 months of ART and more commonly in patients with low baseline CD4 count <50/μL and high pathogen load [6]. It is due to the restoration of pathogen-specific immune responses, which cause immunopathology. It is estimated that 10%–25% of unselected patients who start ART experience this syndrome. It presents either with paradoxical worsening of treated opportunistic diseases or unmasking of a subclinical infection. IRIS may involve many pathogens only some of which affect the GIT. Up to 25% of patients with hepatitis B or C coinfection experience a hepatitis flare or liver enzyme elevation. There may be paradoxical worsening or unmasking of tuberculosis or *Mycobacterium avium* complex (MAC) infection involving the abdomen suggested by fever, abdominal pain, lymphadenopathy, which may be necrotic, and splenic or hepatic microabscesses. Histology reveals granulomata and local culture is frequently positive for the organism. IRIS in coinfection with *Strongyloides* can result in hyperinfection syndrome or worsening of chronic symptoms. ART can be continued unless the IRIS event is severe or life threatening. Corticosteroid therapy is effective in severe tuberculosis-IRIS and corticosteroids or other anti-inflammatory treatment may be of value with other causes of severe manifestations.

CLINICAL APPROACH

Gastrointestinal disease is a common presentation and complaint in people who are immunocompromised due to HIV infection and who are not on ART. In the developed world, there has been a dramatic decline in gastrointestinal opportunistic diseases in those on ART [7]. In endoscopic studies, opportunistic diseases have fallen from 69% to 13% with 80% of patients on ART. Cytomegalovirus infection and esophageal candidiasis have decreased dramatically while noninfectious complaints like gastroesophageal reflux have risen. ART itself can treat conditions like Kaposi's sarcoma, cryptosporidiosis, and microsporidiosis.

The approach to the patient needs to compensate for the fact that clinical symptoms and signs more frequently suggest a syndrome rather than a specific diagnosis or etiology. In the ART era, one can divide patients by ART exposure. Those patients who are ART-naive are more likely to be suffering from an opportunistic disease especially with a CD4 count <100–200/μL or oral candidiasis if CD4 count is not available. More than one infection may account for the symptoms. The patients on ART would be expected to have a drug-related disorder or a non-HIV-related condition similar to the general population. An important caveat is that the ART is effective, and this can be quantified by demonstrating suppression of the viral load <50 copies/ml and a rise in CD4 count from the baseline or nadir CD4 count. If the viral load is suppressed, a CD4 count <200/μL is less likely to have the same significance as in the ART-naïve patient. Tuberculosis is a possible exception to the CD4 count guide as it is more pathogenic than most other opportunistic organisms and occurs at higher counts. There will be people with primary resistance to antiretroviral treatment in whom virologic suppression will not ensue and others who will fail ART after initial virologic suppression. Further patients may be noncompliant. Depending on duration of effective ART or a partial response, these patients will develop clinical failure sooner or later and present as do the group not on ART, unless compliance is improved or a new effective ART regimen is instituted. Those with virologic and clinical failure are likely to increase in number over time. The duration on ART and preexisting illnesses will give an indication of the likelihood of IRIS.

Investigations for opportunistic causes of disease in those who have severe immunosuppression are well validated for the pre-ART era. For those on ART, the prevalence of disease and optimal approach to diagnosis and management is less well defined.

ESOPHAGITIS

In the severely immunocompromised patient, common symptoms are dysphagia and odynophagia. The commonest cause is esophageal candidiasis. Oral candidiasis is usually, but not invariably, present. Treatment is empirically with fluconazole. Other causes of esophagitis in which odynophagia is most prominent include ulcers due to *Cytomegalovirus* (CMV) infection, aphthous or idiopathic ulcers, and herpes simplex. Candida and herpes occur at CD4 counts <200/μL and CMV and idiopathic ulcers at CD4 counts <50/μL. More rarely, the cause can be opportunistic infections including tuberculosis, MAC infection, and histoplamosis, and malignancies including Kaposi's sarcoma, lymphoma and squamous cell carcinoma. In patients not responding to fluconazole (resistant nonalbicans Candida species may be present) or in whom odynophagia is prominent, endoscopy with biopsy (and possibly culture) should be performed. At higher CD4 counts and in people on ART, reflux esophagitis should be considered and presumptive therapy with a proton-pump blocker can be tried for patients with dyspepsia. Other noninfectious disorders such as achalasia and Barrett esophagus are found infrequently, but may become more common in urban settings with more affluent lifestyles.

PEPTIC ULCERATION

Peptic ulceration is less common in HIV-infected populations despite the high prevalence of *Helicobacter pylori* infection. This seems to be a consequence of hypochlorhydria, which seems to be reversed by ART and therefore peptic ulceration may increase in frequency.

DIARRHEA

Diarrhea occurs in more than half of all patients with AIDS. Chronic or persistent diarrhea becomes more common as the CD4 count drops to low levels. With sufficient investigation, the cause can be found in 80% of patients. Chronic diarrhea causes significant morbidity, affects quality of life adversely, and may cause malabsorption of ART and increase the risk of virologic failure and nonadherence.

In the severely immunocompromised patient not on ART, there is a long list of potential opportunistic causes, comprising bacteria and mycobacteria (tuberculosis and MAC), viruses, protozoa, and malignancies (Kaposi's sarcoma and lymphoma). Investigation for the etiology requires stool tests (for MC&S, parasites, and *Clostridium difficile)* and blood cultures for bacteria and mycobacteria. This can be followed by upper GI tract endoscopy with biopsy (for parasitic infections) and flexible sigmoidoscopy or colonoscopy with biopsy (for CMV, malignancy, and mycobacterial disease) where the cause remains inapparent and there is no response to treatment. Intestinal perforation may be caused by CMV colitis, typhoid, or tuberculosis. Treatment for chronic diarrhea is tailored to the specific cause. In areas where investigation will delay therapy or is rationed, it seems reasonable to treat with ciprofloxacin (for bacteria and isosporiasis) and metronidazole (for some protozoa and *C. difficile*) while sending stool for parasites. Albendazole can subsequently be added to cover *Enterocytozoon intestinalis*, one of the treatable microsporidia. There is still no fully effective treatment for cryptosporidiosis and the most common of the microsporidia, *E. bieneusi*. HIV enteropathy, a diagnosis of exclusion, has no specific therapy other than ART. For these causes without specific therapy and the other causes associated with immunosuppression, immune reconstitution by the initiation of ART is essential.

In the ART era, the number of patients experiencing diarrhea has remained static. Opportunistic infections have diminished in importance, while noninfectious causes especially and other infections have become more prominent [8]. Drug-induced diarrhea is the main noninfectious cause, due to the protease inhibitors. Nelfinavir, though infrequently used now, can cause diarrhea in half of those who use it while lopinavir/ritonavir and fosamprenavir/ritonavir cause diarrhea initially in 20%. It appears to be a secretory diarrhea caused by, among others, calcium-dependent chloride conductance and cellular apoptosis, necrosis, and decreased proliferation of intestinal epithelial cells. It is usually of mild-to-moderate severity, develops shortly after starting the drug, and resolves spontaneously. Symptomatic treatment for persistent diarrhea can be attempted, but drug substitution where available is often required, for example with raltegravir or atazanavir/ritonavir. HIV enteropathy may improve with ART, but may also occur in patients receiving ART. The pathogenic mechanisms are not clear but may involve the effect of HIV on the gastrointestinal tract and the gut-associated lymphoid tissue (GALT). Environmental enteropathy and HIV enteropathy have similar morphological and functional characteristics. Autonomic neuropathy and chronic pancreatitis may also contribute to chronic diarrhea. The commonest infectious cause in those on ART is *C. difficile*.

In summary, the investigation of diarrhea will depend on history and examination and duration of diarrhea, modified by CD4 count and viral load and the use of ART. Stool examination may reveal a pathogen. If not, upper and lower tract endoscopy may need to be augmented by radiology for neoplastic lesions and sonography for evidence suggesting mycobacterial disease or lymphoma. ART drugs should be reviewed. If there is still no diagnosis, then HIV enteropathy may be the cause.

HELMINTH INFECTIONS AND HIV

Both helminth infections and HIV significantly alter the immune system [9]. Although the data remain inconclusive, coinfection can influence pathogen establishment, growth, replication and clearance, disease severity, and transmission. Species-specific studies are underway, involving soil transmitted helminths, schistosomiasis, and treatment effects.

HEPATOBILIARY DISEASE

In the ART era, AIDS-related mortality has decreased markedly, so that non-HIV-related causes of death in high-income countries have assumed greater importance. Liver disease accounts for 10%–15% of deaths in these countries, most of which are associated with hepatitis B or hepatitis C viruses [10]. At the clinical level in Africa, the numbers of HIV-infected people on ART presenting with jaundice, hepatitis, or raised liver transaminases have been increasing and often have multiple possible causes and represent a clinical conundrum.

HEPATITIS B VIRUS

Coinfection with hepatitis B (HBV) leads to accelerated progression of HBV. Two drugs in the commonly recommended first-line ART regimens in Africa, tenofovir with either lamivudine or emtricitabine, are effective against HBV. This has led to some countries excluding HBV serology from the ART work-up. In such situations, HBV testing would become necessary in a possible clinical hepatitis flare, or when tenofovir needs to be substituted by dint of renal toxicity or virologic failure. Lamivudine or emtricitabine remain part of most second-line regimens, but resistance of HBV to these agents develops predictably over time. Tenofovir can be continued in a second-line regimen for its HBV effect where deemed appropriate. The lack of HBV testing precludes the opportunity to offer testing and vaccination, if needed, to sexual partners. Coinfection results in an increased risk of hepatotoxicity on ART. Triple therapy ART is used, to prevent HIV resistance developing, in those identified with HBV liver disease requiring HBV treatment but who do not qualify for ART by usual criteria.

HEPATITIS C

The epidemiology of hepatitis C (HCV) in Africa is incompletely understood. HCV testing is not routinely performed in HIV programs except in some tertiary settings and in specific studies.. Coinfection results in accelerated progression of HCV. There are few centers that offer HCV treatment, but this is expected to increase over time, especially with the development of oral drugs for HCV. ART has been clearly shown to be of benefit in coinfected people in reduction of liver-related death. Coinfection increases the risk of hepatotoxicty to ART.

HEPATOCELLULAR CARCINOMA

Hepatocellular carcinoma is a particular concern in coinfection with hepatitis B and hepatitis C. In high-income countries, patients present earlier and with more advanced disease. There are few data from Africa.

CAUSES OF JAUNDICE OR SYMPTOMATIC "HEPATITIS" IN THE PATIENT ON ART

The duration on ART, the antiretroviral drugs used, and the efficacy of ART (based on nadir and current CD4 count and viral load) will inform the differential diagnosis. Comorbid diseases, systemic infections, and other medications will modify this further. There may be non-HIV-related causes such as autoimmune hepatitis. The differential diagnosis includes:

1. Viral hepatitis. Hepatitis B or C flare or progression or acute hepatitis A or B infection. Hepatitis E, Epstein-Barr virus or cytomegalovirus is occasionally the cause.
2. Drug-induced liver injury. This may be a hypersensitivity reaction (Nevirapine, efavirenz, abacavir) or idiosyncratic. Nevirapine is the commonest cause among the antiretrovirals but almost all classes can be implicated. Other common causes are the antituberculosis drugs for both drug-sensitive and multidrug resistant tuberculosis. Cotrimoxazole can cause cholestatic injury, but it can be hepatocellular. Didanosine rarely causes noncirrhotic portal hypertension.
3. IRIS, especially mycobacterial disease. Lymph nodes around the porta hepatis can cause cholestasis.
4. Alcohol-related liver disease.
5. Lactic acidosis with hepatic steatosis.
6. Neoplasm. Non-Hodgkin's lymphoma may present as a primary hepatic condition and may present while on ART. Kaposi's sarcoma is less common.
7. Nonalcoholic fatty liver disease is the hepatic manifestation of the metabolic syndrome [11]. It is more common in those coinfected with hepatitis C. It includes a spectrum from mild fatty liver, to nonalcoholic steatohepatitis (NASH), to cirrhosis. This has been increasing in high-income countries.
8. AIDS cholangiopathy or acalculous cholecystitis. AIDS cholangiopathy occurs in late-stage disease, but may not have been recognized or the patient may be failing ART. It may be caused by chronic infections or neoplasms.

PANCREATITIS

Pancreatitis continues to cause morbidity in the ART era. ART- or HIV-related causes include drugs (didanosine and less commonly stavudine; lamivudine is implicated in children), as a feature of lactic acidosis, as a consequence of hypertriglyceridemia, or due to opportunistic infections, most commonly cytomegalovirus infection.

CONCLUSIONS

The spectrum of GIT illness and the balance between opportunistic diseases and non-AIDS- or HIV-related illness has been shifted by widespread access to ART. A major challenge for the clinician is the development of liver function abnormalities on ART, which can be difficult to elucidate even when liver biopsy is employed. A source of optimism on the horizon is the hope that medication for people with hepatitis C infection will become more affordable and thus widely available.

REFERENCES

[1] Piot P, Quinn TC. Response to the AIDS pandemic – a global health model. N Engl J Med 2013;368:2210–8.
[2] Deeks SG, Lewin SR, Havlir DV. The end of AIDS: HIV infection as a chronic disease. Lancet 2013;382:1525–33.
[3] Sabin CA. Do people with HIV infection have a normal life expectancy in the era of combination antiretroviral therapy? BMC Med 2013;11:251.
[4] Lugassy DM, Farmer BM, Nelson LS. Metabolic and hepatobiliary side effects of antiretroviral therapy (ART). Emerg Med Clin North Am 2010;28:409–19.
[5] Campion EW. Treating millions for HIV – the adherence clubs of Khayelitsha. N Engl J Med 2015;372(4):301–3.
[6] French MA. Immune reconstitution inflammatory syndrome: immune restoration disease 20 years on. Med J Aust 2012;196:318–21.
[7] Wilcox CM, Saag MS. Gastrointestinal complications of HIV infection: changing priorities in the HAART era. Gut 2008;57:861–70.
[8] MacArthur RD, DuPont HL. Etiology and pharmacologic management of noninfectious diarrhea in HIV-infected individuals in the highly active antiretroviral therapy era. Clin Infect Dis 2012;55:860–7.
[9] Weatherhead JE, Woc-Colburn L. Helminth infections and HIV: a double hit. Medscape 2014. October 21.
[10] Allen J, Smith C, Bhagani S. Will antiretroviral therapy reduce HIV-related liver risk? Curr Opin HIV AIDS 2014;9:48–53.
[11] Kelly P, Saloojee H, Chen JY, Chung RT. Noncommunicable diseases in HIV infection in low- and middle-income countries: gastrointestinal, hepatic and nutritional aspects. J Acquir Immune Defic Syndr 2014;67:S79–86.

CHAPTER

12

CHARACTERISTICS OF INFLAMMATORY BOWEL DISEASE IN SUB-SAHARAN AFRICA

Isidor Segal

G.I.T. Unit, Chris Hani Baragwanath Hospital and University of Witwatersrand, Soweto, South Africa

IBD IN SSA

Inflammatory bowel disease (IBD) is uncommon in Black populations of SSA but is increasing, particularly ulcerative colitis (UC). This is contrary to the global situation in which IBD is prevalent in the HICs of Europe, the USA, and increasing in Asia [1].

UC IN SSA

UC is infrequent in SSA. In a study from Soweto (Johannesburg), only 18 cases were reported prior to 1975 [2,3]. A later study [4](2014) from the S.A. Cancer Registry indicated the rise of I.B.D from 1970 to 2014—74 patients with UC were diagnosed.

In LIMCs, the overwhelming number of patients with bloody diarrhea may obscure the diagnosis of UC. In order to rule out other causes of bloody diarrhea, the GIT Unit at Baragwanath Hospital excluded the following causes: amoebiasis, shigellosis, lymphogranuloma venereum, tuberculosis (TB), bilharziasis, Yersinia enterocolytica, and Crohn's disease (CD). With regard to the various discriminatory tests, patients were initially treated with metronidazole (this, despite negative tests for amoebiasis).

In SSA, it is important to exclude these diseases before considering a diagnosis of UC. UC was confirmed on the basis of radiological, endoscopic, and histological evidence.

The grading of the severity of the colitis was based on the classification of Truelove and Witts [5]. The results indicated that UC was more common in women and the mean age of onset was 36 years. The diagnosis was made at a very late stage—approximately 3 years after the onset of symptoms. Patients were anemic—67% had severe iron deficiency anemia. All patients had chronic bloody diarrhea and weight loss was present in most. 89% had moderately severe or severe disease. 72% had either total colitis or extensive colitis. There were no deaths during hospital stay and no patients had toxic megacolon or bowel perforation. The bias toward a severe presentation may be because patients with mild disease do not come to hospital for treatment. Most of the patients did not attend the follow-up clinic except for acute attacks. In view of the possible high risk of colorectal cancer in this group and the unwillingness to return for follow up, it may be advisable to offer patients with extensive disease early surgery.

Digestive Diseases in Sub-Saharan Africa. https://doi.org/10.1016/B978-0-12-815677-3.00012-8

UC appears to be increasing, albeit slowly in SSA. In other countries of SSA, the uncommonness of the disease has been confirmed. In Nigeria, three cases of UC and one of CD were reported in 2011 in a 3-year period [6]. There have also been case reports of IBD from Sudan, Cote d'Ivoire, Senegal, Ethiopia, and Zimbabwe. However, recent reports from South Africa show that the number of new IBD cases (*all races*) diagnosed annually from 1970 to 2009 has risen from 12 patients to 87 patients annually [4].

The incidence and characteristics of UC is similar in the White South African population to that of HICs populations.

CROHN'S DISEASE IN SSA

CD is rare in Black populations of SSA. The Colored (mixed race) population, however, has a high incidence of aggressive, complicated IBD, and in whom CD is more prevalent than UC [4]. It is in the Colored population that a significant rise in CD has occurred. In a report of a cross-sectional study of all consecutive CD patients seen at 2 large IBD referral centers in the Western Cape, South Africa, between September 2011 and January 2013, 194 CD subjects were identified: 152 (78%) were Colored; 35 (18%) White; and 7 (4%) were Black [4]. Coloreds were found to have aggressive CD with high rates of TB among IBD patients [4].

In a report on the global incidence of CD, the incidence in South Africa is noted as 1–3 per 100,000 people [7]. Only seven patients with CD were reported prior to 1975. Since then, in 6 years, only 12 cases have been diagnosed at Baragwanath Hospital. The S.A.IBD Registry showed that the number of Black people with CD was 29 for the period 1970–2014 [4]. The experience at Baragwanath Hospital is of particular significance as it probably reflects the pattern of CD in the Black populations of SSA.

The diagnosis of CD is essentially one of exclusion. It is then confirmed on the basis of clinical, radiological, endoscopic, and histological evidence. Thus TB, Yersinia enterocolitis, amoebiasis, lymphogranuloma venereum, and bilharziasis were all considered before the diagnosis was confirmed. In addition, a 3–6 month trial of anti-TB therapy was administered before definitively diagnosing CD. TB responds dramatically to anti-TB therapy.

Validity of the data

The apparent rarity of CD and UC in LMICs could be attributed to the following: (1) lack of medical personnel and special diagnostic facilities, (2) difficulty in maintaining continuity in the care of the individual patient. It is true that follow up in the setting of SSA is difficult, but special care was taken to follow up these patients because of the rarity of these diseases and to obtain clarity as to the diagnosis, (3) the sketchy and incomplete nature of medical records, (4) the relative frequency of other causes of bloody diarrhea thus particularly masking UC, and (5) under-reporting of the diseases. It is asserted therefore that the low prevalence of the diseases in LMICs may simply be a reflection of the scarcity of medical resources or confusion of the diseases with the infective dysenteries. In other words, the diseases may be inadvertently missed. This may be true, but in large well-equipped centers of excellence, it is probable that patients with these diseases are likely to be diagnosed. It is therefore reasonable to conclude that CD and UC are uncommon entities in Black populations of SSA.

In relationship to the clinical, radiological, and sometimes even the histological appearances of CD and TB, it is our practice to administer a 6-week to 3-month course of anti-TB therapy in all cases in whom the diagnosis of CD is considered. The response to treatment, if the diagnosis is TB, is dramatic. However, one can see in the following reservations regarding the difficulties that may arise in diagnosis.

THE ASSOCIATION OF CD AND TB [8,9]

South Africa has the second highest incidence of TB in the world (948/100,000 a year) after Swaziland. 615 IBD patients were included in a study by Deetlefs et al. [8]. Coloreds comprised 394 (64%) of the study group and Blacks 31 (5%).

Evidence of TB was found in 11.7% of IBD patients either before or after a diagnosis of IBD. The high rate of previous TB and positive association with race likely reflects the high burden of TB in a socioeconomically disadvantaged community.

Epstein et al. [9] noted that in LMICs TB complicates both the diagnosis and management of CD. In addition, empiric treatment may delay treatment for CD or make it difficult to confirm or refute a diagnosis of intestinal TB at a later stage. The difficulties with regard to radiological, endoscopic, and pathological differentiation are emphasized as are the challenges in the treatment of CD in TB endemic regions. In this regard, consideration must be given to the fact that steroids, immunological, and biologicals may be disastrous if administered to TB patients. Caution is advised that before a diagnosis of CD is made to actively exclude intestinal TB on clinical, radiological, and endoscopic evidence. Culture for *M. tuberculosis* is mandatory in all patients bearing in mind that histology also has its pitfalls. It is important to actively assess patients in LIMCs, who are on treatment for CD so that reactivation of latent TB does not occur.

CONCLUSIONS WITH REGARD TO IBD IN SSA

CD and UC are uncommon in SSA Black populations but appear to be increasing. The low incidence may be due to under-reporting or not diagnosing the condition because of the large numbers of people in SSA who have bloody diarrhea. This refers particularly to UC. Before establishing a diagnosis of UC, it is important to exclude other diseases causing bloody diarrhea. With regard to CD, a high index of suspicion is required to exclude abdominal TB, which can mimic CD. The difficulties of diagnosis and treatment of CD in a region where TB is endemic are emphasized, as are the pitfalls in treating CD.

Centers of excellence and resources are necessary for diagnosis, treatment, and follow up of the patients. In seeking a possible cause/s for the paucity of IBD, it is proposed that in the context of poverty in infancy an adverse environmental situation conditions the gut and paradoxically, acts as a protective factor against subsequent digestive diseases in adulthood [10]. It has also been proposed that the Hygiene hypothesis [11,12] may be a factor in this low prevalence. The hypothesis recognizes that the "reduced microbial exposure" concept applies to a much broader range of chronic inflammatory diseases than asthma and hay fever, which includes diseases such as type 1 diabetes and multiple sclerosis, some types of depression and cancer. The hygiene hypothesis also proposes that several chronic inflammatory diseases including IBS are increasing in prevalence in LMICs because a changing environment has perturbed immunoregularity circuits, which normally terminate inflammatory responses.

ASPECTS OF IBD IN HICS

IBD comprises two main subtypes—UC and CD. They are chronic, lifelong immune-mediated diseases characterized by acute flares and in some cases by persistent chronic symptoms. The inflammatory response in UC affects the mucosa and submucosa, whereas in CD the inflammatory response extends from the intestinal mucosa to serosa.

Ten to fifteen percent are termed indeterminate colitis because it is not possible to categorize them as either UC or CD until the diseases manifest into definite entities [13]. Also a change in diagnosis occurs in a small proportion of cases over time as the disease evolves [1]. The patterns of family aggregation suggest that there is a genetic basis to both diseases [13].

ULCERATIVE COLITIS

Although the etiology is unknown, there is evidence that genetic, immunological, and environmental factors contribute to UC. There is a relationship between UC and bacterial flora with an increased number of adherent *Bacteroides* spp. and *Enterobacteriaceae* spp. present in inflamed bowel segments.

EPIDEMIOLOGY

UC frequently affects young people between the ages of 15 and 25 years and in some series there is a second peak between 55 and 65 years of age although it can affect individuals of any age [13]. There appears to be a slight male predominance in some HICs for UC. Since the Second World War, UC incidence in European and North American countries are beginning to level off, while rising incidence rates have been noted in traditionally low prevalence regions such as in Eastern Europe, South America, and Asia [1].

In HICs of Europe, first-degree relatives of patients with UC have a 10- to 15-fold risk of developing the disease.

Smoking is an established risk factor, which increases the risk and worsens the clinical course of CD, but is protective in UC. Nonselective antiinflammatory drugs (NSAIDs) appear to exacerbate the clinical activity of IBD. Studies on the role of appendectomy, diet (including refined sugars), drugs, seasonal variation, water supply, social circumstances, hygiene factors, and vaccination have been largely controversial and inconclusive [1].

CLINICAL SIGNS AND SYMPTOMS

The cardinal symptom of UC is bloody diarrhea or the passage of blood and mucus. Nocturnal diarrhea is common and the associated symptoms include fever, abdominal pain, which may be either left lower quadrant or rectal pain, abdominal cramps, urgency, and tenesmus. Weight loss may or may not be present depending on how much of the colon is involved. Most guidelines employ a cut off of “more than 6 weeks” of diarrhea to differentiate IBD from infectious diarrhea [1]. Systemic features include fatigue, malaise, arthralgia, fever, and weight loss. Extraintestinal manifestations such as arthritis, episcleritis, and erythema nodosum and pyoderma gangrenosum may complicate presentation in 10% of cases, and may rarely precede bowel symptoms [1].

UC is classified as proctitis (rectal involvement only), left-sided (involving sigmoid colon with or without involvement of descending colon) or extensive colitis (involvement, which extends beyond the splenic flexure) based on findings on endoscopy. The disease has a high morbidity. The extent of disease involvement is important as it influences the form of treatment as well as the frequency of colorectal cancer surveillance [1].

At presentation, about 40% of adult UC cases have proctitis, 40% have left-sided colitis, and up to 20% suffer from extensive colitis. Progression of disease extends over time. The disease is divided, according to severity of symptoms, clinical findings, and laboratory values into mild, moderately severe, and severe disease [1].

One out of every four UC patients will require colectomy after 10 years of disease [1].

TREATMENT

The principles of management are to assess clinical activity, location and extent of disease, treatment of acute disease, improve and maintain patient's quality of life, correct nutritional deficiencies. It is important to maintain steroid-free remissions and to prevent complications, hospitalizations, and surgery [14].

MANAGEMENT OF IBD IN HICS: PROBLEMS IN LMICS

NARROWING THE GAP

IBD has been found to adversely affect quality of life with negative impact on psychological, familial, social, and occupational dimensions of life [15].

IBD affects more than 2.5 million people in Europe and is becoming increasingly frequent in Asia and in developing countries [16]. In the USA, it is currently estimated that about 1–1.3 million people have IBD [17]..

In HICs, the facilities for training medical personnel and the technical equipment necessary for the diagnosis of IBD are readily available. In addition, the expertise required for assessing, managing, and follow up of patients is of a high standard. The team required for this task includes record keeping, specialists, doctors in training, pathologists, radiologists, nurses, social workers, and psychotherapists. This manpower is not available in LICs. The question arises as to how can the gap can be narrowed with the available manpower in LICs. Also under-reporting may be a crucial factor in the uncommonness of IBD.

Initially an infrastructure should be established. A start has been made by setting up training institutes in Africa with the assistance of the World Gastroenterology Organization. These institutes encourage interchange and training of doctors and includes training in record keeping, clinical training, and endoscopy. However, more interventions are required and the assistance of HICs would be valuable. The establishment of libraries, donations of journals, computers, and endoscopic instruments (even if used) would be encouraging. In addition, international co-operative studies between HICs and LMICs would have collateral benefits.

REFERENCES

[1] Thia K, Ooi O. Changing patterns of inflammatory bowel disease in a global context (ulcerative colitis). In: Gastroenterology and hepatology manual. A clinicians guide to a global phenomenon. Sydney: McGraw-Hill; 2011. p. 48–65. [Chapter 6].

[2] Segal I. Ulcerative colitis in a developing country of Africa: the Baragwanath experience of the first 46 patients. Int J Color Dis 1988;3(4):222–5.

[3] Segal I. Intestinal tuberculosis, Crohn's disease and ulcerative colitis in an urban black population. S Afr Med J 1984;65:37–44. 14 Jan.

[4] Epstein D. Inflammatory bowel disease in Cape Town. In: Experience and evidence. 2014. [Abstract SAGES].
[5] Truelove SC, Witts LJ. Cortisone in ulcerative colitis. Br Med J 1955;2:1041–8.
[6] Ukwenya AY, Ahmed A, Odige VI, et al. Inflammatory bowel disease in Nigerians. Still a rare diagnosis? Ann Afr Med 2011;10(2):175–1795.
[7] World Gastroenterology Organisation. Global guideline: Inflammatory bowel disease: a global perspective. WGO Publication; June, 2009.
[8] Deetlefs E., Watermeyer G., Seggie R. et al. A descriptive study of tuberculosis and inflammatory bowel disease cohort from Cape Town, South Africa, 2009, SAGES Congress.
[9] Epstein D, Watermeyer G, Kirsch RR. Review article: the diagnosis and management of Crohn's disease in populations with high-risk rates for tuberculosis. Aliment Pharmacol Ther 2007;25:1373–88.
[10] Segal I, Walker ARP, Wadee A. Persistent low prevalence of western digestive diseases in Africa: confounding aetiological factors. Gut 2001;48:730–2.
[11] Strachan D. Family size, infection and atopy: the first decade of the 'hygiene hypothesis'. Thorax 2000;55(90001):2S. https://doi.org/10.1136/thorax.55.suppl_1.S2.
[12] Hygiene Hypothesis. Living with nature, http://www.hygienehypothesis.com/, p 2.
[13] Stenson WF. Inflammatory bowel disease. In: Yamada T, Alpers DH, Owyang C, et al., editors. Textbook of gastroenterology. 2nd ed. Philadelphia: Lippincott Company; 1995. p. 1748–808.
[14] Australian Guidelines for General Practitioners and Physicians. Inflammatory bowel disease. 4th ed. Gastroenterological Society of Australia Management; 2017. p. 11. [DRAFT].
[15] Israeli B, Graff LA, Clara I, et al. Low prevalence of disability among patients with inflammatory bowel diseases a decade after diagnosis. Clin Gastroenterol Hepatol 2014;12:1330–7.
[16] Kappelman MD, Rifas-Shiman SL, Kleinman K, et al. The prevalence and geographic distribution of Crohn's disease and ulcerative colitis in the United States. Clin Gastroenterol Hepatol 2007;5:1424–9.
[17] Loftus Jr. EV. Clinical epidemiology of inflammatory bowel disease: incidence, prevalence, and environmental influences. Gastroenterology 2004;126:1504–17.

CHAPTER

ABDOMINAL TUBERCULOSIS IN SUB-SAHARAN AFRICA (SSA)

13

M.K. Kgomo*, O. Mwantembe†

Head of Gastroenterology Unit, University of Pretoria, Pretoria, South Africa Former Head of Gastroenterology, Steve Biko Academic Hospital, University of Pretoria, Pretoria, South Africa†*

World Health Organisation (WHO) publications in 2010 reported that worldwide, 8–10 million people get infected by TB every year. In 2003, it was reported that the worldwide death rate from tuberculosis (TB) was more than 6000 people per day grossing 3 million per year.

More than 72% of patients with TB are from South Asia and Sub-Saharan Africa (SSA) [1].

TB of the abdomen (ATB) is the sixth most frequent extrapulmonary site [2].

Figures may be higher because of difficulty in diagnosing ATB in poorly resourced countries of SSA. It is not surprising that FDG-PET scan images show a more extensive involvement by TB mycobacterial in HIV-positive patients than clinically suggested [3]. The key message is that the condition is underdiagnosed and undermanaged.

HIV and TB are more or less synonymous and hence justify the combination of specialized clinics for treatment. The annual risk of developing active TB, when coinfected with HIV, is 20–30 times the risk in seronegative individuals [3].

Approximately 80% of patients with TB in South Africa are HIV positive [2], and of these 40%–70% of TB involve an extra pulmonary site [4].

Only 15%–20% of ATB patients have concomitant active pulmonary TB, making the diagnosis of ATB a specific challenge. It entails another approach, which may be unconventional.

SSA has the highest rates of active TB per capita, driven by the HIV epidemic. In 2010, about 68% of all people living with HIV resided in SSA-a region with only 12% of global population [1].

SSA bears the brunt of being the region with the highest burden of both HIV infection and TB. Although only 10% of immunocompetent individuals infected with TB develop active disease in their lifetime, 50% of those coinfected with HIV develop active TB [1]. South Africa and Nigeria are listed among the 22 highest countries in the world with TB and the leading two in SSA. The nine leading SSA with the highest prevalence of TB are in decreasing order: South Africa, Democratic Republic of Congo, Ethiopia, Nigeria, Kenya, Mozambique, Zimbabwe, Tanzania, and Uganda [1]. Progress in the management and containment of these conditions is imperative [5] (Figs. 13.1 and 13.2).

The chest is the commonest system affected by TB. It is important to realize that 15%–25% of pulmonary TB patients [6] also have ATB, which is the sixth most common extra pulmonary manifestation, and comprises 3%–4% of the 17% extra pulmonary TB.

The challenge is to realize that this great mimicker may be affecting the gastrointestinal system especially when a patient presents with a few subtle symptoms and signs [4, 7]. Even though TB is a chronic disease, it can present acutely for example as perforation in the abdomen.

Digestive Diseases in Sub-Saharan Africa. https://doi.org/10.1016/B978-0-12-815677-3.00013-X

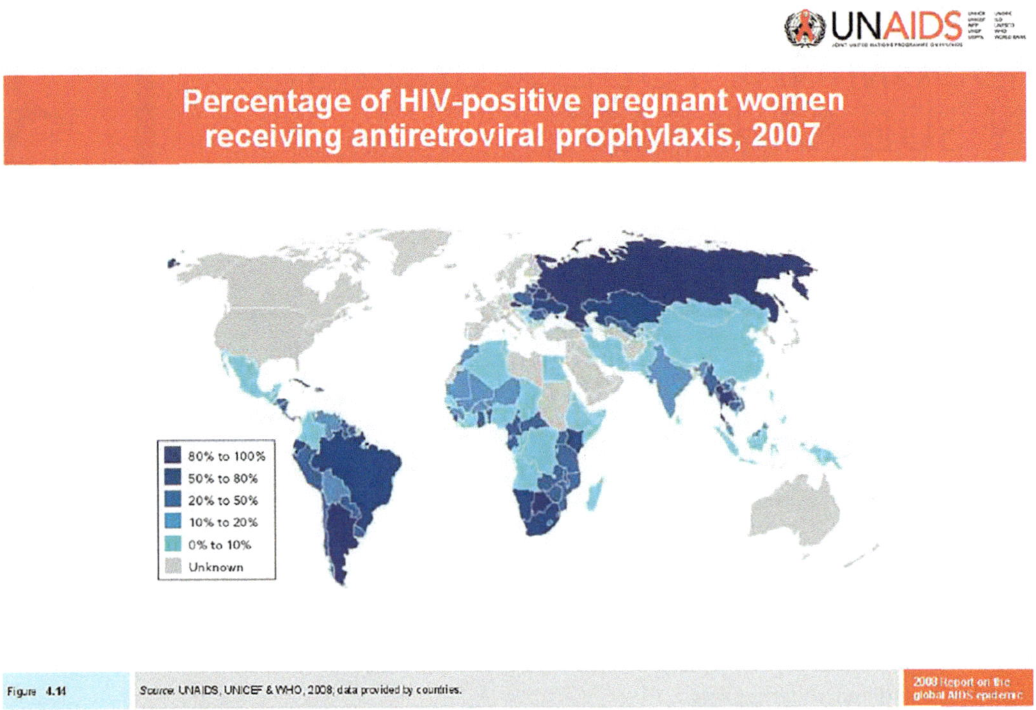

FIG. 13.1

Percentage of HIV positive pregnant women receiving antiretroviral prophylaxis.

Source: UNAIDS, UNISEF & WHO, 2008.

GASTROINTESTINAL INVOLVEMENT

TB can affect any part of the gastrointestinal system. The involvement is conventionally classified as follows:

1. Intestinal by far the commonest site
2. Peritoneal 4%–10%
   Wet with ascites—commonest
   Plastic with adhesions
   Dry with fibrosis with omental thickening and loculated ascites
3. Lymphadenopathy
4. Visceral TB involving the solid organs 15%–20% [8]

PERITONEUM

The spread to this area is hematogenous and lymphogenous. Classically it shows tubercles on laparoscopy [9]. This may be associated with an exudative ascites. The ascites results from secretions from the tubercles.

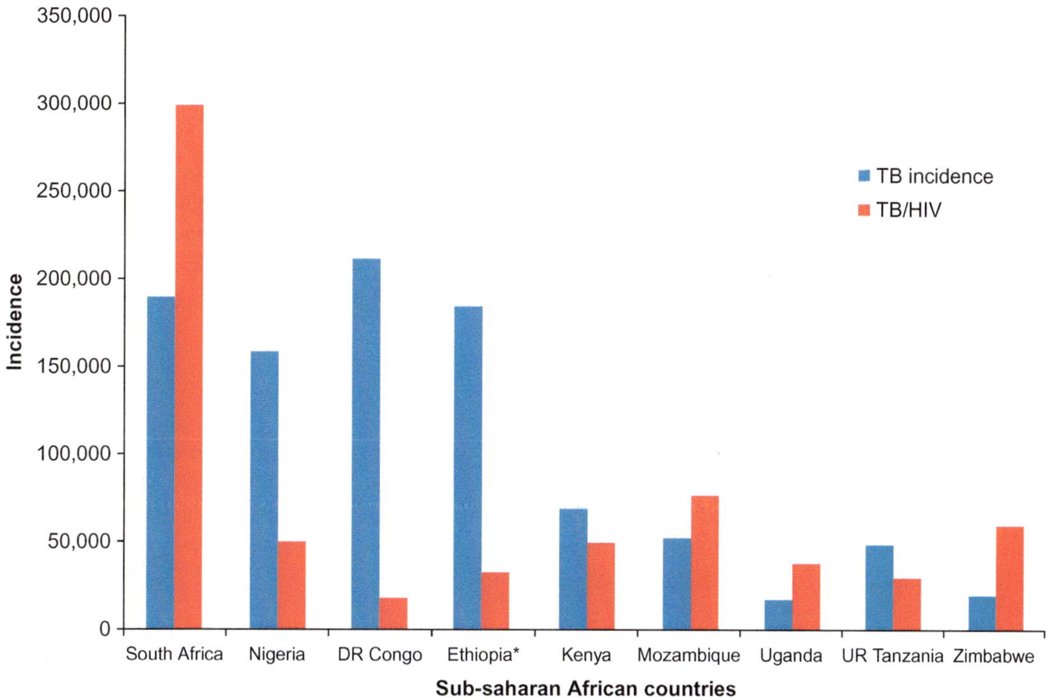

FIG. 13.2

Tuberculosis with HIV in Subsaharan Africa.

Reproduced from WHO's global tuberculosis report 2011.

INTESTINAL

This is the commonest area of involvement and the ileocecal region tends to be the most involved. ATB affects regions of the intestines in the following order of frequency ileocecal, jejunum/ileum, colon, anorectal, stomach, appendix, duodenum, and esophagus [10]. The ileocecal region is most frequently affected because of the slow transit, the abundance of lymphocytes in Peyer's patches and the increased absorption in this area. It is not uncommon, as in Crohn's disease (CD), that a diagnosis is only made when a patient has undergone laparotomy for appendicitis.

The pathologic characteristics include [11]:

1. Ulceration
2. Hypertrophy
3. Ulcerohypertrophic
4. Ulceroconstrictive
5. Fibrotic

These also determine the clinical presentation.

In the tubular part of the gastrointestinal system, TB can lead to a number of effects, which may show different presentations. The edema caused by inflammation can result in intestinal obstruction

[12] as can the fibrosis [13] that may follow successful treatment. The inflammation may cause matting of the intestines with its sequelae. In other parts of the body, it may cause perforation.

LYMPHADENOPATHY

This is the second most commonly affected system of the abdomen. It accounts for 25% of ATB and mainly afflicts young HIV-positive adults [14]. The large size of these nodes may cause obstruction in the adjacent intestinal lumen. The lymph nodes most commonly affected are the para aortic and mesenteric, tabes mesenterica. This could be because the nodes drain from the most frequently diseased parts of the intestines [11]. This is unlike the Hodgkin's and non-Hodgkin's lymphoma that tend to affect lower abdomen and pelvic lymph nodes.

OTHER INTESTINAL ORGANS

The spleen, the kidneys, and the pancreas may be involved—the spleen more often than the kidney or the pancreas. In clinical practice, the latter is usually thought of as neoplasm before TB is considered.

Most commonly, the disease affects several parts of the intestine but to a different degree. The commonest affliction is a combination of intestinal and lymph node involvement [13]. It is more often that the lymph nodes are involved without the tubular involvement.

CLINICAL PRESENTATION

There are no diagnostic clinical presentations or signs. The most important method of arriving at a diagnosis of this condition with protean manifestation is to think of the possibility of ATB in the appropriate setting. The diagnosis should be considered in patients with pulmonary TB or recently treated pulmonary TB or indeed people whose immunity may be compromised. In addition, there is an increasing number of de novo ATB without any obvious predisposing factors [15]. The predisposing factors include diabetes, chronic renal failure on CAPD, patients on chemotherapy or immunotherapy such as antitumor necrosis therapy, and indeed the genetics of the mycobacterium [16]. It is also justifiable to think of ATB as a natural extension of pulmonary TB, as previous to when effective treatment was available, 55%–90% of patients who succumbed to pulmonary TB had ATB as well on post mortem [17].

Patients are often asymptomatic, especially those who are immune compromised.

Most commonly patients present with constitutional symptoms, but these symptoms are hardly specific. Symptoms include generalized malaise, fever, night sweats, loss of weight, and fatigue.

Although there may be specific findings attributable to specific organs on laboratory investigations, seldom are there any florid organ-specific symptoms.

On clinical examination the patients may be wasted, have some mild pallor and features that may be in keeping with their underlying condition.

The chest may show signs of pulmonary involvement, but this is not invariable.

On examination of the abdomen, tenderness may be present but not prominently rebound as in acute bacterial peritonitis. If there is rebound, then perforation should be considered. The doughy feeling classically described is rare and this results from the plastic type of peritoneal disease [18].

In the absence of ascites quite frequently there is organomegaly, hepatomegaly, and splenomegaly. When ascites is present, the usual features of shifting dullness or/and fluid thrill may be elicited.

INVESTIGATIONS

The gold standard for diagnosing TB is identifying the acid fast bacilli. This is seldom achieved with abdominal TB (3%).

As a general rule, especially in resource-challenged areas where abdominal TB is endemic, the following guidelines should be followed, as described by Lingenfelder et al. [19]

1. Clinical signs suggestive of ATB
2. Radiological features in keeping with ATB
3. Histopathology or microbiology in keeping with TB, or/and
4. A therapeutic response to treatment

CLINICAL SIGNS

The cardinal feature in the diagnosis of ATB is a high index of suspicion as the symptoms and signs are not specific. The constitutional symptoms stated here should alert one to the possibility of ATB in the diagnosis.

INTESTINAL ATB

The commonest involvement is the ileocecal region. There may be localized tenderness in the right iliac fossa with a palpable mass. The differential diagnosis neoplasm, appendicular mass, amoebae or gastrointestinal histoplasmosis, or other fungal infections.

The classical doughy feeling of the abdomen described in literature is very rare and it is indeed subjective [18].

Visible abdominal borborygmi and signs of intestinal obstruction may result from any of the pathological processes of ATB.

THE ASCITES OF ATB

Ascites tends to be straw colored with a protein content of 3g/dl or more. However, what may be more helpful is the blood-ascites gradient as this corrects for a patient who may be hypoalbuminemic. A blood—ascites albumin gradient of less than 1.1 g/dL is in keeping with an exudate. The serum glucose ascites ratio is less than 0.96. The WBC ranges from 150 to 4000/cc and is predominantly above 70% lymphocytes.

Microbiology is seldom positive for AFB usually less than 3%. Cultures of the ascitic fluid have a slightly better yield of approximately 20%.

Assay of adenosine deaminase (ADA) levels levels aids the diagnosis—ADA deaminates adenosine to inosine in the lymphocytes. ADA level in ascitic fluid above 33 U/L has a sensitivity of 100%, specificity of 97%, and accuracy of 98% in diagnosis ascitic TB [20]. Similarly an Ascitic—blood ADA ratio above 0.985 also strongly supports TB as the cause of ascites [21].

However, in the presence of HIV, all these tests may not be very helpful as they depend largely on an intact cell-mediated immunity.

PERITONEAL INVOLVEMENT

The typical signs of an acute peritonitis are rare. Rebound tenderness is usually mild and severe unless intestinal perforation is present.

In the wet variety of peritonitis, signs of ascites such as shifting dullness are positive. The finding of tense ascites is not typical of peritoneal TB unless underlying liver cirrhosis is present. Sonar guided or direct tapping of the ascites should be part of the physical examination.

LYMPHADENOPATHY

Lymph node involvement accounts for 25% of ATB and is found commonly in a younger HIV-positive group.

The lymph nodes may rarely be palpable but more commonly they may be observed by ultrasonography (USG). The nodes involved are typically the upper abdominal nodes. This group includes para-aortic-, peri-pancreatic-, mesenteric-, and periportal lymph nodes [10]. This reflects the commonly involved areas of the intestine, which drains to these nodes. The nodes in ATB tend to be matted. This distribution is in contrast to the Hodgkin's and non-Hodgkin's lymphomas where the lower pelvic nodes are more commonly involved.

VISCERAL SOLID ORGAN INVOLVEMENT

The liver is the second most commonly affected abdominal solid organ (the genitourinary tract is the commonest site). Other terms used for liver involvement are atypical TB, tubercular hepatitis, tubercular cholangitis, and serohepatic TB [22].

The spleen is a major site of endothelial activity. Thus, hepatosplenomegaly is a feature of ATB. However, it is not invariable as these organs are only affected in less than 10% of ATB due to military TB.

Pancreatic TB commonly results from contiguous or hematogenous spread. It is uncommon accounting for only about 4.7% of ATB [23]. This may present as colonic perforation, abdominal abscess, portal vein thrombosis, obstructive cholangiopathy, or as low-grade chronic abdominal pain. With imaging it may mimic pancreatic carcinoma. It differs from carcinoma in that there is no ductal involvement as it is a parenchymal disease unlike carcinoma of the pancreas and there is no vascular involvement. Besides as it is not the commonest manifestation of ATB, usually other features of ATB are present such as lymphadenopathy. It may however finally require image-guided percutaneous or endoscopic fine-needle biopsy for a histological diagnosis.

LABORATORY INVESTIGATIONS

Usually these are nonspecific. Anemia of chronic disorders, high ESR, low albumin, and slightly raised transaminases may be present. If there is significant involvement of the liver, the alkaline phosphatase and gamma GT may be markedly raised and this will mimic mechanical cholestasis.

ULTRASONOGRAPHY

This is one of the most informative investigations in resource-constrained environments. A well-performed USG may yield useful information and may be pathognomic of ATB. USG is informative as to lymph node, organ, and peritoneal disease involvement.

INTESTINES

Matted loops with more than 15-mm thickened walls. The most marked intestinal wall thickening is around the ileocecal region and this area should be targeted.

Clubbed sandwich or sliced bread sign, which is due to a collection of ascites between radially oriented bowel loops may also be present [24].

LYMPHADENOATHY

Matted enlarged nodes in the para-aortic, mesenteric, and prepancreatic area may show caseation and calcification—a combination that increases the likelihood of ATB. The combination illustrates a collection of nodes with hypoechoic centers.

Calcification in the lymph nodes in the absence of a primary tumor increases the likelihood of ATB.

PERITONEUM

The peritoneum may be thickened but uniformly, so unlike the patchy thickening associated with peritoneal metastases.

The ascites may be lobulated, with fine septae and echogenic debris.

SOLID ORGANS

These may be enlarged and may show characteristics such as multiple abscesses in the spleen.

RADIOLOGY

Chest X-ray is only positive for TB in less than 25% of ATB patients.

Plain Abdominal X-ray may be informative as it may show intestinal obstruction with areas of calcification.

Barium meal may show diffuse mucosal irregularity with narrowing, which may be more pronounced in the ileocecal region with a string sign as in CD.

Supportive signs include:

Rapid emptying of barium from the small intestines (Sterlin's sign)

Flocculation and segmentation—features of malabsorption

BARIUM ENEMA

This may show any of the following:

Fleischner's sign or reversed umbrella sign an appearance resulting from a thickened widened triangular ileocecal valve and a widened cecum with a straight ileum.

There may also be signs of bowel irritability as shown by rapid transit of barium in the involved areas—Sterlin's sign.

CT SCAN

CT Scan is the best radiological test. The features covered become more prominent and not subject to interpreter or varying sonographic capabilities. The ascites can be further analyzed by measuring the Hansfield, which is usually between 25 and 45 in tuberculous ascites.

The lymph nodes can be better visualized and if they are in the usually involved areas and show peripheral rim enhancement. This supports the diagnosis of ATB.

ENDOSCOPY

Colonoscopy may show hyperemic friable mucosa, cobblestoning, and pseudopolyps. The features are more prominent in the ileum so as to make terminal ileoscopy very important. Biopsies from this area should be sent for histology as well as microbiology.

The features in the ileum may be very similar to CD. However, there are a number of subtle differences (discussed later).

LAPAROSCOPY

This is arguably the most rewarding examination although the most invasive [25]. It is particularly risky in the setting of TB where adhesions and loops of bowel may be matted and intestines adherent to the peritoneum.

Laparoscopy may show:

A thickened peritoneum with yellow tubercles, which are amenable to biopsy. This is pathognomonic and gives a high yield of 65%–70% confirming the diagnosis of ATB.

In fact, the finding of fibrotic strands or adhesions within a turbid ascites is strongly suggestive of TB.

Inspection of the organs usually shows them to be studded with tubercles.

FDG-PET scan in HIV-negative patients suffering from TB shows more extensive involvement by mycobacterial TB as compared to contrast CT, suggesting that TB involvement was underestimated prior to the use of this test [3].

MRI

The advantage of MRI is that no radiation is involved. There is still debate as to whether this adds much toward the diagnosis. MRI is not as sensitive as CT in showing calcification.

MICROBIOLOGY

The disease tends to be paucibacillar and the yield even on culture is less than 20%. As regards direct microscopy of ascitic fluid, a yield of less than 3% is the most that can be expected.

HISTOPATHOLOGY

Histopathology findings include Ulceration, nonspecific granulation tissue with numerous polymorphs and microabscesses—60%

Noncaseating granulomata—35%

Characteristic caseation with Langhans giant cells in only 25% of specimens.

25% show nonspecific variable changes.

Immunohistochemistry with special stains increases the yield.

Pathognomic giant cells are uncommonly observed as is the AAFB in ATB. It becomes important therefore that another modality should be tried for the ante-mortem diagnosis. This is where a therapeutic trial is not only diagnostic but can be life saving.

Is the diagnosis CD or intestinal tuberculosis (ATB)?

Since the first description of what was CD by Dalziel in 1769, the question has been asked as to whether CD, as it later was known, was not TB or a form of TB. It is now clearly established that these are different entities. Whereas the finding of acid fast bacilli and caseating granulomata is diagnostic of TB, this is a very rare finding in ATB the frequency being as low as 5%. The matter has become more relevant as CD once more or less confined to the HICs has become more common (though still rare) in LMICs. There has also been a substantial migration of populations from LMICs to HMICs. Relevant to compounding, this is HIV where any form of TB including pulmonary is paucibacillar.

The corollary has also become a problem. This is where ITB has been misdiagnosed as CD in the western world with delay in proper treatment and dire consequences from prolonged wrong treatment. In China up to 65% of ITB has been misdiagnosed as CD [26].

Both CD and intestinal tuberculosis tend to have a similar distribution as regards which parts of the intestines are primarily affected.

A therapeutic trial with antituberculous drugs would have been the answer but for the fact that we now have organisms that are multidrug resistant and the drugs are not without side effects. Therapeutic trials have sometimes gone on for two years before reconsideration. This again does not augur well for the long-suffering patient.

There are however features that may help distinguish the two conditions and these have to be taken in the context of the entire clinical picture.

For example, in the absence of any other explanation, the presence of tubercles on the peritoneum with ascites [27] and matted intestinal lymphadenopathy would strongly point toward intestinal tuberculosis rather than CD. Generally, the peritoneal involvement would support ITB while perianal involvement would support CD [28].

The entire question has to be approached as a one big picture from the history to the latest investigations so that one can distinguish the Zebra from The striped Horse.

Though the history is markedly similar, a history of more than 12 months of persistent diarrhea with hematochezia points toward CD other than ITB [29]. The latter has a history of usually less than

7 months and fever is more prominent but hematochezia very rare. It must be added that because of the longer Crohn's history, failure to thrive is also a prominent feature.

On clinical examination, it is the extra intestinal manifestations of mainly CD when present that may guide to a diagnosis. These include pyoderma gangrenosum, arthritis, uveitis, and others. However, these are not invariable in CD and indeed not common enough to make their absence a factor in excluding CD. What is of note is the presence of perianal disease. Perianal disease aids support to the diagnosis of CD [28]. When the finds on clinical examination include ascites, lymphadenopathy, and a high swinging fever above 38.5 Celsius, one would be inclined to opt for ITB other than CD [30]. Again these more or less classical presentations are exceptions other than the norm.

The distinguishing characteristics of the ITB ascitic tap have been covered [20, 31].

Endoscopy adds a lot to differentiation. Though the distribution of ileocolonic involvement is the similar the pattern is different. The CD involvement tends to be longitudinal, longer, and deeper. On ileoscopy a spared ileal valve but affected ileum favors more CD than ITB whereas a patulous ileocecal valve with transverse ulcers favors a diagnosis of ITB [32]. The appearance of "fronds" on the mucosa, a result of high granuloma activity, favors TB other than CD.

Abdominal sonar as earlier the finding of matted intestinal lymphadenopathy in the lower abdominal lymph nodes points toward TB [33]. This in the presence of ascitis is more or less diagnostic of ITB.

By far the most informative modality is the abdominal CT scan. Thankfully CT is now increasingly becoming available in poorer parts of the world. CT features of abdominal TB have been described here. There are also features that favor other than excluding CD such as "The Comb sign" [34]. MRI, not available in most resource-constrained areas, has yet to prove useful in this scenario. This is "jejunization of the ileum"—the hypervascularization of the mesentery around the ileum. The linear densities on the mesenteric side of the ileum formed by the fibrofatty proliferation and perivascular inflammatory infiltration give an appearance of a comb on CT imaging.

Serology in general has proved disappointing as there is a lot of overlap in the results [35–37]. However, the Gold quantiferon test has proved useful in the setting of BCG vaccines as none of its targets overlap with the antigens of BCG. The limitation is that it just shows present or previous contact with TB. This is useful in the before instituting Anti- TNF therapy in CD but it does not add much to diagnosing active disease.

The histological features of a typical Langhans TB tubercle are clear especially with caseation. The CD granuloma are small, less than 400um, and less than 4 per high power field [38]. And the latter do not caseate.

Distinguishing CD from ATB can be very difficult and a therapeutic trial of anti-TB therapy may be necessary. Response to therapy in ATB is swift with marked improvement after 2-3 weeks.

TREATMENT

The usual drugs administered are Isoniazid (INH), Ethambutol, *Pyrazinamide, and Rifampicin for the first two months followed by* rifampicin and INH. It must be emphasized that as long they are on INH, pyridoxine must be given to avoid neuropathy.

There are different recommendations regarding the duration of therapy. ATB should have a total 9–12 m treatment, but some studies have shown that the 6 m used for thoracic TB gives similar outcomes [39–41].

It is important that the full course of therapy is completed. This results in resolution of the disease and also avoids the emergence of resistant strains. Those that have multidrug resistant (MDR) and extreme drug resistant (XDR) TB should be managed in Regional Centers.

Constitutional symptoms usually start abating within 24–48 h. However, obstructive symptoms that may be due to lymphadenopathy may exacerbate initially as the nodes tend to swell on starting therapy. However, they decrease in size overtime. The consensus is to avoid steroids as they may predispose to perforations.

SUMMARY

TB has always been a disease with severe morbidity and mortality. In its recent history, it has undergone three phases. Initially before the 1946 discovery of effective therapy, TB had a very high mortality. It is noteworthy that approximately 80%–90% patients who died from pulmonary TBs had evidence of ATB. This was most likely due to seeding from swallowed sputum. With the advent of effective therapy and an improvement in the standards of living, the disease became less common and less ferocious. However, with the advent of HIV, it has resurged not only in the affected but also in the immunocompetent patients, as there is a higher chance of seeding from the contact to the affected. Also because of the prevalence and sometimes inadequate treatment in terms of time and dosage, there has been an increase in treatment resistant TB.

ATB is fairly common in the immunocompromised patient and as discussed here has no pathognomonic signs or symptoms. The diagnosis is usually made by having a high index of suspicion ATB has protean manifestations [42]. A therapeutic trial may be the only way forward not only to make a diagnosis but also to save lives.

Advances in diagnostic methodology would greatly assist in arriving at a diagnosis timeously.

REFERENCES

[1] Mukhtar A, Adeiza A. HIV-associated tuberculosis: a Sub-Saharan African perspective. Sub-Saharan J Med 2014;1(1):1–14.

[2] Islam DL. Lessons from emergency laparotomy for abdominal tuberculosis in the HIV/AIDS era. S Afr J Surg 2014;52(1):10–2.

[3] Sathekge MM, Maes A, Kgomo M, et al. Fluorodeoxyglucose uptake by lymph nodes of HIV patients is inversely related to CD4 count. Nucl Med Commun 2010;31:137–40.

[4] Golden MP. Extrapulmonary tuberculosis: an overview. Am Fam Physician 2005;72:1761–8.

[5] World Health Organisation. Global tuberculosis report. Geneva: WHO; 2014. http://apps.who.int/iris/bitstream/10665/137094/1/9789241564809_eng.pdf.

[6] Sharma SK, Mohan A. Extra-pulmonary tuberculosis. Indian J Med Res 2004;14:316–53.

[7] Sinkala ESG. Clinical and ultrasonographic features of abdominal tuberculosis in HIV positive adults in Zambia. BioMed Central 2009;44(9):471.

[8] Tan KK, Chen K, Sim R. The spectrum of abdominal tuberculosis in a developed country: a single institution's experience over seven years. J Gastrointest Surg 2009;13:142–7.

[9] Rai S, Thomas M. Diagnosis of abdominal tuberculosis; the importance of laparoscopy. J R Soc Med 2003;96:586–8.

[10] Marshall JB. Tuberculosis of the gastrointestinal tract and the peritoneum. Am J Gastroenterol 1993;88(7):989–99.
[11] Marks IN. Abdominal tuberculosis baillierers. Clin Trop Med Common Dis 1998;3:329.
[12] Hovath D, Whelan RL. Intestinal tuberculosis: return an old disease. Am J Gastroenterol 1998;93:692–6.
[13] Aston NO. Abdominal tuberculosis. World J Surg 1997;21:495–9.
[14] Suri S, Gupta S, Sori R. Computed tomography in abdominal tuberculosis. Br J Radiol 2014;72:853.
[15] Sterling TR, Dorma SE, Chaisson RE, et al. Human immuno deficiency sero-negative adults with extra pulmonary tuberculosis have abnormal innate immune responses. Clin Infect Dis 2001;33(7):976–82.
[16] Yang Z, Yang D, Kong Y, et al. Clinical relevance of Mycobacterium tuberculosis plcD gene mutations. Am J Respir Crit Care Med 2005;171:1436–42.
[17] Haddad FS, Ghossain A, Sawaya E, Nelson AR. Abdominal tuberculosis. Dis Colon Rectum 1987;30:724–35.
[18] Al-Quoran AA, Satti MB, Al-Freihi HM, et al. Abdominal tuberculosis in Saudi Arabia: a clinico-pathological study of 65 cases. Am J Gastroenterol 1993;88:75.
[19] Lingenfelser T, Zak J, Marks IN, et al. Abdominal tuberculosis; still a potentially lethal disease. Am J Gastroenterol 1933;88:744.
[20] Dwivedi M, Misra SP, Misra V, Kumar R. Value of ADA estimation in diagnosis of tubercular ascites. Am J Gastroenterol 1990;85:1123–5.
[21] Bhargava DK, Shriniwas, Chopra P. Peritoneal tuberculosis; laparoscopic pattern and its diagnostic accuracy. Am J Gastroenterol 1992;87:109–11.
[22] Pereira JM, Madureira AJ, Vieira A, Ramos I. Abdominal tuberculosis: imaging features. Eur J Radiol 2005;55:173–80.
[23] Veerabadran P, Sasnur P, Subramanian S, Mappagounder S. Pancreatic tuberculosis—abdominal presenting as a pancreatic abscess and colonic perforation. World J Gastroenterol 2007;13(3):478–9.
[24] Jain R, Sahney S, Bhargava DK, Berry M. Diagnosis of abdominal tuberculosis: sonographic findings in patients with early disease. Am J Roentgenol 1995;165(6):1391–5.
[25] Mohamed A, Bhat N, Abhukhater M, Riaz M. Role of laparoscopy in diagnosis of abdominal tuberculosis. Int J Infect Dis 2009;8:1–6.
[26] Tonghua L, Guozong P, Minzhang C. Crohn's disease: clinicalpathologic manifestations and differential diagnosis from enterocolonic tuberculosis. Chin Med J 1981;94:431–40.
[27] Epstein D, Watermeyer G, Kirsch R. Review article: the diagnosis and management of Crohn's disease in populations with high risk for tuberculosis. Aliment Pharmacol Ther 2007;25:1373–88.
[28] Palmer KR, Patil DH, Basran GS, Riodarn GF, Silk DB. Abdominal tuberculosis in urban Britain—a common disease. Gut 1985;26:1295–305.
[29] Amarupak DN, Patel ND, Rane PS. Diagnosis of Crohn's disease in India where tuberculosis is widely prevalent. World J Gastroenterol 2008;14(5):741–6.
[30] Marshall JB. Tuberculosis of the gastrointestinal tract and peritoneum. Am J Gastroenterol 1993;88:989–99.
[31] Sharma M, Bhatia V. Abdominal tuberculosis. Indian J Med Res 2004;120(4):305–15.
[32] Pulimood AB, Amarapukar N, Ghoshal U, et al. Differentiation of Crohn's disease from intestinal tuberculosis in India in 2010. World J Gastroenterol 2011;17(4):433–43.
[33] Jain R, Sawhney S, Bhargava DK, Berry M. Diagnosis of abdominal tuberculosis. Sonographic findings in patients with early disease. AJR Am J Roentgenol 1995;165(6):1301–5.
[34] Meyers MA, McGuire PV. Spiral CT demonstration of hypervascularity in Crohn's disease "vascular jejunisation of the ileum" or the "Comb sign". Abdom Imaging 1885;20:327–32.
[35] Patel N, Amarapurkar DN, Kulshrestha PP, Agal S, Pramanik BR, Gupta P. Utility of serological markers in inflammatory bowel disease. Indian J Gastroenterol 2002;21:A37.
[36] Makharia GK, Sachdev V, Gupta R, Lal S, Pandey RM. Anti-*Saccharomyces cerevisiae* antibody does not differentiate between Crohn's disease and intestinal tuberculosis. Dig Dis Sci 2007;52:33–9.

[37] Patel N, Amarapurkar DN, Kulshrestha PP, Agal S, Pramanik BR, Gupta P. Utility of serological markers in inflammatory bowel disease. Indian J Gastroenterol 2002;21:A37.
[38] Pulimood AB, Ramakrishna BS, Kurian G, Peter S, Patra S, Mathan VI, Mathan MM. Endoscopic mucosal biopsies are useful in distinguishing granulomatous colitis due to Crohn's disease from tuberculosis. Gut 1999;45:537–41.
[39] Balasubramanian R, Nagarajan M, Balambal R, et al. Randomised controlled clinical trial of short course chemotherapy in abdominal tuberculosis: a five-year report. Int J Tuberc Lung Dis 1997;1:44–51.
[40] Marshall JB. Tuberculosis of the gastrointestinal tract and peritoneum. Am J Gastroenterol 1993;88(7):989–99.
[41] Kapoor VK. Abodominal tuberculosis. Postgrad Med J 1998;74(874):459–67.
[42] Segal I, Tim LO, Mirwis J, et al. Pitfalls in the diagnosis of gastrointestinal tuberculosis. Am J Gastroenterol 1981;75:30–5.

CHAPTER

COLORECTAL CANCER IN SUB-SAHARAN AFRICA

14

T.E. Madiba*, F. Makumbi†

*G I Cancer and Research Centre, University of Kwa-Zulu, Durban, Natal, South Africa**

Specialist Physician-Gastroenterologist, Natal, South Africa†

EPIDEMIOLOGY

Colorectal carcinoma (CRC) is the third most common cancer and is responsible for 1.2 million new cases per year worldwide and over 600,000 deaths annually [1,2]. It is the third most common cause of cancer-related death after lung and stomach cancer [1]. Colorectal cancer occurs with great variability in the different parts of the world and it is not uniformly distributed in all population groups [3]. Incidence rates are nearly fourfold higher in developed countries where it accounts for 9.4% compared with underdeveloped countries where it accounts for 2.5% of all malignancies [4–6]. The age standardized incidence is 21–45.7 per 100,000 in developed countries and ranges from 3.3 to 7.2 per 100,000 in Africa [1].

Concerns do exist, however, regarding the systematic under-reporting of CRC in African countries due to the fact that access to medical care is not readily available with many patients not reaching treatment [5]. Despite these concerns, there is evidence that CRC is increasingly being diagnosed in African countries [4,7,8].

When discussing diseases such as CRC with great variability, discussion of population groups becomes inevitable. It therefore becomes important to define these population groups since their definitions differ from those in the rest of the world. The term African refers to black people of indigenous African origin. Indians are of Asian origin and Whites are of European origin, and represent groups of recent seaborne immigrants to the continent that make up the South African population. In Southern Africa, the term "Colored" is an ethnic label that refers to people of mixed ethnic origin who possess ancestry from Europe, Asia, and various Khoisan and indigenous African tribes of Southern Africa.

Incidence rates of CRC are higher in North Africa than in sub-Saharan Africa [4]. In countries like South Africa where African and White communities are found, the frequency of colorectal carcinoma is much lower among Africans compared to non-Africans especially Whites [4,9–11].

The age at presentation for CRC in African series is 35–54 years [11–16]. This contrasts with the 66–70 years reported in international literature [11,12,16]. Patients presenting before the age of 40 years are reported to be 1%–5% worldwide [16,17] and about 14%–54% in African countries [11–14]. In South Africa where Blacks and Whites coexist, the mean age at presentation and the age distribution for Whites mirrors that in the world literature whereas Black Africans present a decade earlier [15]; accordingly, 18%–36% of Black Africans present at age ≤40 years [12–15] compared to 2%–5% of Whites [11].

Digestive Diseases in Sub-Saharan Africa. https://doi.org/10.1016/B978-0-12-815677-3.00014-1

ETIOLOGY

DIET, ENVIRONMENTAL, AND LIFESTYLE FACTORS

High rates of CRC are seen in societies with western-type lifestyle such as diet rich in red meat and animal fat but lacking in vegetables, fruit, and other dietary fiber [6,8,18–20]. The typical African diet consists of a carbohydrate-based bolus type of meal, which cannot be consumed alone but requires liquids like soup [19]. High-fiber diet increases the speed of intestinal transit so that any carcinogen in the diet and feces will be in contact with the mucosa for a short time [18]. Furthermore, vegetable and fruit fiber tends to provide better protection than cereal-type fiber [19].

In sub-Saharan Africa, there has been an increase in health risk behaviors associated with CRC, such as smoking, decreased physical activity, alcohol consumption, and poor diet with resultant increase of the health-related consequences such as obesity, diabetes, and hypertension [5,6]. In some sub-Saharan countries, changes in exposure to environmental factors have resulted in the African population overtaking Whites in the prevalence of obesity, hypertension, and diabetes [4]. Moreover, evidence exists that migrants who have adopted the dietary customs of their new country tend to assume the colorectal cancer risk of their adopted country [11,19]. The established geographical variation and the migratory evidence confirms the suggested role played by environmental factors and lifestyle changes in the development of colorectal cancer [11].

PRECURSOR LESIONS

Etiological risk factors for colorectal cancer include colorectal polyps, inflammatory bowel disease, and family history of colorectal cancer [3]. The two common types of polyps are adenomatous and hyperplastic polyps. Whereas previous studies have suggested the rarity of polyps and inflammatory bowel disease among Black Africans [4,11,13,21,22], more recent work has demonstrated similar mucosal proliferation patterns in Africans and Whites and have postulated a more rapid conversion of polyps to carcinoma among African patients [16].

The risk for CRC increases with increasing number and size of adenomatous polyps. The propensity of adenomatous polyps larger than 1 cm to develop CRC is 1%–1.5% per annum [2]. Villous histology and polyps with sessile stalks are associated with higher malignant potential than tubular histology and pedunculated stalks, respectively.

Colorectal cancers arise from dysplastic adenomatous polyps via an adenoma-carcinoma sequence [2]. The most common pathway, occurring in 70%–85%, is the Chromosomal Instability Pathway in which a cascade of genetic mutations occurs involving the inactivation of tumor suppressor genes such as p53, and suppression of DNA repair together with simultaneous activation of oncogenes such as the Adenoma Polyposis coli (APC) gene. This leads to progressive accumulation of multiple genetic mutations, which results in the transition from normal mucosa to benign adenoma to severe dysplasia to frank carcinoma [2].

Although hyperplastic polyps have previously been presumed to have little or no association with CRC, recent evidence suggests that such an association does exist [2], the risk factors being large polyp size (>1 cm diameter), a focus of adenoma within the polyp (mixed hyperplastic-adenomatous polyp), >20 hyperplastic polyps in the colon, a family history of hyperplastic polyposis, and a family history of CRC [2]. Hyperplastic polyps are linked to CRC via the recently reclassified serrated adenoma, which arises within a hyperplastic polyp but differs from it by abnormal proliferation of crypt epithelium and nuclear atypia [2]. The serrated adenoma transforms into colorectal cancer via a BRAF genetic mutation and extensive DNA methylation, which tends to inactivate mismatch repair genes [2]. This pathway is seen in 15%–20% of colorectal cancers [2].

Syndromic colorectal cancer is a result of germ line mutations of specific genes [2]. The familial adenomatous polyposis (FAP) syndrome is the result of the APC gene mutation and is inherited as a Mendelian autosomal dominant gene [2]. Patients develop hundreds or thousands of adenomatous polyps throughout the colon beginning after puberty and inevitably develop colorectal cancer [2]. In hereditary nonpolyposis colorectal cancer (HNPCC), affected patients typically have a few polyps and the cancer typically occurs in the right colon in middle age [2]. The Amsterdam II criteria are used to clinically diagnose HNPCC. There is sparse literature on these inherited conditions in Africa except for isolated areas in the west coast of South Africa.

PATHOLOGY

The main type of cancer affecting the colon and rectum is adenocarcinoma, which occurs in about 95%–98% of colorectal cancer. Whereas the geographical variation of the anatomical distribution of colorectal cancer is now established, there is no agreement with regard to the predominance of left- vs right-sided colonic tumors [4]. The picture differs from that in African-Americans in whom proximal colonic tumors tend to predominate in comparison to their White counterparts [6]. Cancer of the rectum is more common than cancer of the colon in African series, occurring in 51%–67% of reported cases [8,21], in contrast to the 30%–31% seen in high-incidence populations [4].

Differentiation in CRC is based on the degree of preservation of normal glandular architecture. Approximately 20% of CRC are poorly differentiated and have a poor prognosis [2]. About 15% are classified as mucinous, or colloid, because of prominent intracellular accumulation of mucin; the signet-ring variant of mucinous carcinoma is very aggressive and carries a poor prognosis [2].

Colorectal cancer spread is by contiguous, lymphatic, hematogenous, and transcelomic routes in descending order of frequency. Like all gastrointestinal cancers contiguous spread can be lateral or longitudinal in the intramural plane. Carcinoma in situ, or high-grade dysplasia, denotes cancer that is confined to the mucosa without penetration of the muscularis mucosa [2]. It is highly unlikely to produce metastases because the lymphatic and vascular channels are below the muscularis mucosa [2].

As shown in Table 14.1, Stage I disease occurs in 1%–15%. Approximately 20%–25% of CRC patients present with synchronous distant metastases at initial diagnosis, and about 25%–50% of patients without metastases at presentation develop metachronous distant metastases [23]. Tumor staging in African series have yielded conflicting results with some suggesting advanced stage of disease presentation in African patients [8,12] while others show a picture similar to that in the western literature [15].

Table 14.1 The Staging of Colorectal Cancer in Various Studies [2,23,29]

Stage	%
Stage I	1%–15%
Stage II	20%–30%
Stage III	30%–40%
Stage IV	20%–25%

Up to 30% of patients with colorectal cancer present with complications, the most common being obstruction (8%–29%) and perforation (3%–8%) and fistula into other hollow visceral organs occurs in 1% [15]. A similar picture pertains in sub-Saharan Africa [15].

CLINICAL FEATURES

Key clinical features of colorectal cancer are alteration in bowel habit (mainly constipation) and bleeding from the rectum. Abdominal pain may be present if there are associated complications such as perforation and obstruction and a mass may rarely be palpable.

Chronic blood loss from the right colon may lead to anemia. There is deterioration in the general health and loss of weight and appetite. Vague right iliac fossa pain may be present and a mass may be palpable. Obstructive symptoms are uncommon and may be associated only with large tumors.

As the left colon is narrower and has more solid contents, obstruction is relatively common. Increasing constipation and alternating bouts of diarrhea are characteristic. Distension and colicky pain are common and blood and mucus are often passed per rectum.

Bleeding is prominent feature of rectal carcinoma. The presence of a tumor in the rectum causes persistent bowel symptoms. A classical feature is early morning diarrhea. The patient experiences the desire to defecate but passes only a little slime and blood, because fluid stool passes more easily through the obstruction than solid stool, thus giving the impression of diarrhea (spurious diarrhea). Other features include urgency, tenesmus, and a feeling of incomplete defecation. The general health may remain good. Whereas chronic bleeding from carcinoma is the rule, serious life-threatening hemorrhage rarely occurs [3].

DIAGNOSIS

Rectal examination and sigmoidoscopy are essential since almost half of all colorectal cancers lie in the rectum and one-third lie within reach of a finger at rectal examination [24].

Complete colonoscopy should be performed in all patients prior to surgical resection, to identify synchronous tumors and to remove adenomatous polyps. If complete visualization of the colon prior to surgery is not possible, colonic evaluation should be performed within 6 months of surgery [24]. Although double contrast barium enema allows evaluation of the entire colon, its diagnostic sensitivity is inferior to colonoscopy and it does not allow biopsy [25]. A colonoscopy will be necessary after this investigation.

Rectal cancer requires additional preoperative local imaging to establish the T and N stages of the tumor and the best preoperative staging modalities are magnetic resonance imaging (MRI) and Endorectal ultrasound (ERUS), both of which are superior to Computerized Tomography (CT) [2].

Preoperative serum testing of carcino-embryonic antigen (CEA) is useful for prognosis and as a baseline for comparison with postoperative levels [2]. An elevated serum level preoperatively is a poor prognostic indicator [2].

Routine metastatic work-up involves looking for metastases in the liver and chest, the standard modality being CT chest and abdomen with an accuracy of approximately 85% for large (>1 cm) lesions [2,24]. Chest radiographs and liver ultrasound are relatively noninvasive, inexpensive, and have a low false-positive rate and can be used for the detection of lung and liver metastases in resource limited settings [24]. Whereas liver-specific MRI is more accurate than CT in detecting focal liver metastases, it is not routinely used [2].

TREATMENT

The resection of colorectal cancer is the mainstay of treatment and is guided by similar oncologic principles [26]. Based on the knowledge of locoregional tumor spread, wide resection of the primary lesion *en bloc* with the draining lymph node basin, leaving a clear margin of clearance is the standard procedure [26].

The demonstration of microscopic invasion of the submucosa for a distance from the primary tumor has necessitated a margin of 10 cm proximally and 5 cm distally to reduce the incidence of local recurrence [26]. Since recent evidence suggests a minimal propensity for rectal cancer to spread >1 cm distally [26], a 2-cm distal margin is now accepted as satisfactory in the rectum [26]. The higher incidence of local recurrence associated with the compromise of the radial (circumferential) margin [26] has led to the description of total mesorectal excision along defined anatomic planes to achieve safe margins and, subsequently, acceptable local control and survival rates [26]. Wide resection of the distal colon, rectum, mesorectum, and, when necessary, the anus is important to minimize local recurrences [26]. While early intramucosal rectal lesions are amendable to local resection, they are generally uncommon in African series [26].

ADJUNCTIVE THERAPY

Adjunctive therapy involves the use of chemotherapy with or without radiotherapy in addition to or in place of surgery. The intended purpose of adjuvant (postoperative) therapy is to cure those patients who are rendered free of disease by operation but who will ultimately have a relapse without further intervention due to undetected microscopic metastases present at the time of operation [26]. Patients without such microscopic metastases are cured by operation alone, and if they could be identified accurately, they would require no further therapy [26]. Because of the differences in the natural history of colon and rectal cancer, adjunctive treatment strategies have evolved in separate ways for patients with rectal cancer and those with colon cancer [26].

The need for postsurgical treatment for colon carcinoma is dictated primarily by the stage of the cancer. It is recommended that patients with stage III disease should receive adjuvant chemotherapy. Neoadjuvant therapy is rarely indicated for colonic cancer. Unlike tumors located above the peritoneal reflection, primary tumors below the peritoneal reflection have a tendency to invade through the bowel and mesorectum making adjuvant therapy necessary [26].

In patients with clinically resectable rectal carcinoma, radiation therapy decreases the risk of local recurrence and increases the chance for sphincter preservation [26]. In patients with locally advanced or unresectable disease, radiation therapy has the added role of potentially increasing the resectability rate [26]. The role of chemotherapy, regardless of the resectability status, is to enhance the effectiveness of radiation (radiosensitization) and to improve survival by eradicating microscopic distant metastases [26].

PROGNOSIS

Complete surgical resection is associated with the best local control and disease-free survival. Although 80% of patients with a colorectal malignancy initially present with locally advanced disease that is amenable to curative surgical resection, up to 40% suffer local, regional, or systemic recurrence as well as metachronous metastases [24].

Differences in the natural history after operation of rectal cancer versus colon cancer have been demonstrated [26]. Whereas local failure is a significant clinical problem in rectal carcinoma, this failure pattern is less frequent after operation for colonic carcinoma [26].

Relapse after resection for colon cancer occurs in 20%–40% of patients after curative surgery [26]. Factors in favor of recurrence are transmural invasion, lymph node metastases, degree of differentiation, presence of lymphovascular and perineurial invasion, margins of resection, and the presence of complications such as obstruction and perforation at the initial operation [26].

In rectal cancer, the most important cause of local failure is residual disease left in the perirectal soft tissues after resection [26]. The circumferential or radial margin is often in the range of millimeters because it is difficult to remove a large amount of perirectal fat along with the primary specimen [26].

FOLLOW-UP

The primary goal of surveillance following curative resection of colorectal cancer is to identify recurrent disease at an asymptomatic stage when it is amenable to re-resection and cure. The secondary goal is the early identification of new polyps or cancers.

History and physical examination of asymptomatic patients, although routinely done, rarely identify a resectable recurrence [24]. In many cases, recurrences are discovered through investigation of symptoms that occur between scheduled follow-up visits [24]. This is because many patients wait until their next scheduled appointment before reporting symptoms, enabling a period of time to elapse that could theoretically allow resectable disease to become unresectable [24].

FOLLOW-UP INVESTIGATIONS

After apparently complete colorectal cancer resection, the serum CEA level almost always normalizes. Failure to normalize postoperatively suggests incomplete resection. A sustained and progressive rise after postoperative normalization strongly suggests cancer recurrence [2]. CEA is the first abnormal test in 38%–66% of patients who develop a recurrence, with an estimated lead time of 4–6 months [24]. An abnormal CEA level mandates prompt surveillance colonoscopy to exclude local recurrence and abdominal imaging to exclude metastases [2].

Regular chest radiographs and abdominal ultrasound are essential initial follow-up investigations [24]. CT is the modality of choice for the detection and confirmation of liver metastasis and other intra-abdominal or pelvic recurrences [2].

Regular colonoscopy should be performed at 2–3 yearly intervals. However, since most local recurrences are the result of residual disease, they start extraluminally and cannot be detected endoscopically until they are very advanced [24] and can be missed at colonoscopy.

SCREENING

Colorectal cancer screening is the testing of asymptomatic patients for the presence of premalignant conditions such as polyps. Approximately 30% of individuals harbor risk factors for CRC including premalignant conditions and family or personal history of CRC [25]. The other 70% of individuals are considered average risk. Colon and rectal cancer can be prevented through screening and removal of premalignant polyps. Average-risk individuals should be offered screening beginning at age 50 years and high-risk individuals should start screening earlier [25].

A screening test must be able to detect the majority of premalignant or incident cancers at the time of testing and should achieve acceptable levels of sensitivity and specificity [27]. The choice of modality for CRC screening along with the associated risks and benefits must be discussed between the practitioner and the individual patient [25].

Available screening methods include (i) stool-based screening methods, (ii) double contrast barium enema, (iii) flexible sigmoidoscopy, (iv) colonoscopy, and (v) CT Colonography. Stool-based tests include fecal occult blood testing (FOBT), fecal immunohistochemistry test (FIT), and stool DNA testing. Stool-based tests are by themselves not likely to prevent cancer since a positive test needs confirmation by colonoscopy and are not necessary if colonoscopy screening is available.

The disadvantage of these noninvasive tests is that they need to be repeated at regular intervals to be effective; and, if the test is abnormal, colonoscopy will be needed [27]. FOBT reacts to the peroxidase activity of hem. The test is therefore liable to reaction with other peroxidases in the feces (certain fruits, vegetables, and red meat). Dietary restrictions are therefore necessary to avoid false-positive results.

Double contrast barium enema examines the whole colon and rectum, is easily accessible, and has a lower complication rate (Fig. 14.1). However, it requires full bowel preparation and does not allow removal or biopsy of lesions seen.

Flexible sigmoidoscopy examines the rectum and left colon up to the splenic flexure and can detect 80% of colorectal cancers. It is more expensive and has a much higher yield than the rigid instrument. Its limitations are inability to visualize the area beyond the reach of the sigmoidoscope, and, where CRC is diagnosed at sigmoidoscopy, a colonoscopy is still necessary to rule out synchronous lesions [25].

The standard screening test is colonoscopy (Fig. 14.2). It is able to detect synchronous lesions, anastomotic or local recurrences, and metachronous lesions [24]. However, it is costly, invasive and requires bowel preparation.

Virtual colonoscopy or CT colonography is an evolving technique [25] and, like DCBE, it has a disadvantage of not allowing histological evaluation and colonoscopy becomes necessary after virtual colposcopy.

CHALLENGES FOR SUB-SAHARAN AFRICA

In contrast to developed countries, most developing countries like sub-Saharan Africa have a very low risk for CRC. These countries have limited heath resources, making screening and surveillance difficult. As such extensive diagnostic tests are not feasible in these countries [5]. Consequently, flexible sigmoidoscopy and DCBE may be the only investigations available in cases where cost is a major factor and colonoscopy is therefore not readily accessible.

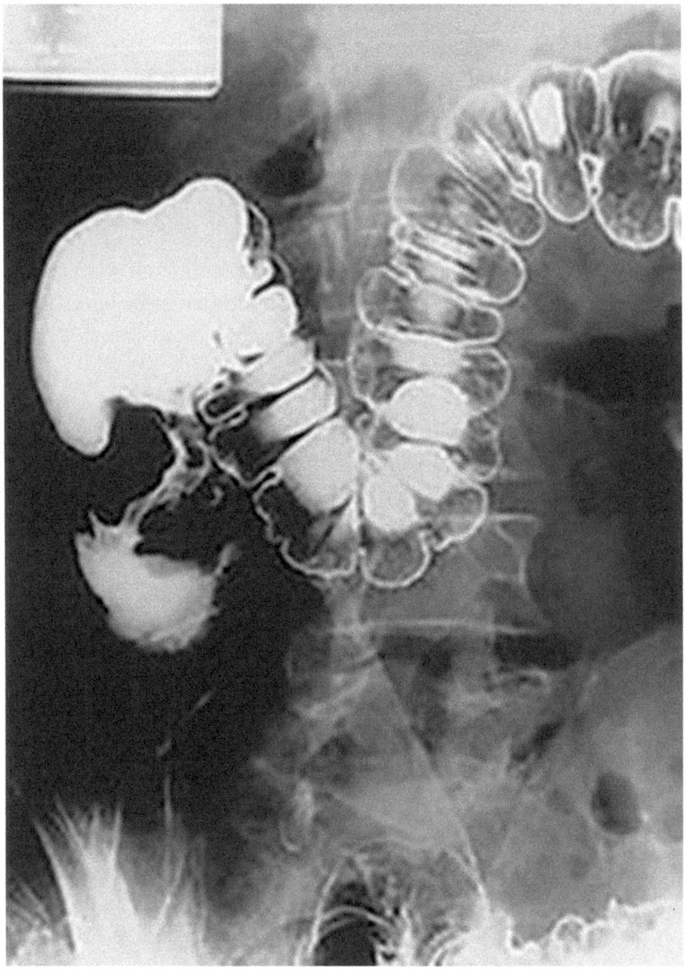

FIG. 14.1

Barium enema showing an "apple-core" lesion in the sigmoid colon.

Since tumors tend to present at an advanced stage of the disease, local excision of rectal lesions using trans-anal endoscopic surgery is not feasible in sub-Saharan Africa. The lower age at presentations for African patents makes a recommendation for commencement of screening at a lower age than whites attractive and this would be in keeping with the recommendation that African Americans begin screening at an earlier age than whites based on their younger age of presentation [6]. The changing subsite prevalence and its association with age and race have now prompted discussions regarding the appropriateness of flexible sigmoidoscopy, previously touted as a screening tool. There is argument now for colonoscopy as the procedure of choice [8,28].

In sub-Saharan Africa, CT and MRI are not always readily available, and chest x-ray and abdominal ultrasound are the only investigations available. Furthermore, a large shortfall still exists for basic radiation services and much work is needed to keep pace with the burgeoning populations of many African countries [5].

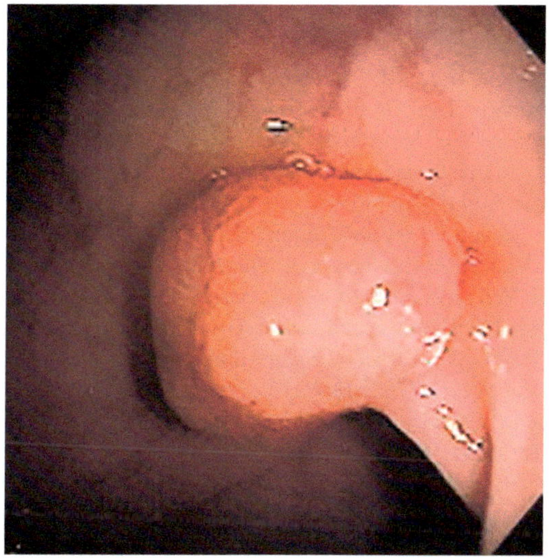

FIG. 14.2

Colonoscopy picture shows malignant lesions.

REFERENCES

[1] Jemal A, Bray F, Center M, Ferlay J, Ward E, Forman D. Global cancer statistics. CA Cancer J Clin 2011;61(2):69–90.

[2] Cappell MS. Pathophysiology, clinical presentation, and management of colon cancer. Gastroenterol Clin N Am 2008;37(1):1–24.

[3] Madiba T. Colorectal cancer. In: Adeloye A, Adekunle O, Awojobi O, editors. Davey's companion to surgery in Africa. Eruwa, Nigeria: Acecool; 2009.

[4] Parkin D, Ferlay J, Hamdi-Chérif M, et al. Cancer in Africa. Lyon: IARC Scientific Publication; 2003.

[5] Graham A, Davies Adeloye LG, Theodoratou E, Campbell H. Estimating the incidence of colorectal cancer in Sub-Saharan Africa: a systematic analysis. Glob Health 2012;2(2):1–14.

[6] Kauh J, Brawley O, Berger M. Racial disparities in colorectal cancer. Curr Probl Cancer 2007;31:123–33.

[7] Béjar LM, Gili M, Infantes B, Marcott PF. Effects of changes in dietary habits on colorectal cancer incidence in twenty countries from four continents during the period 1971-2002. Rev Esp Enferm Dig 2011;103(10):519.

[8] Saidi H, Nyaim EO, Githaiga JW, Karuri D. CRC surgery trends in Kenya, 1993–2005. World J Surg 2008;32(2):217–23.

[9] Segal I, Edwards CA, Walker AR. Continuing low colon cancer incidence in African populations. Am J Gastroenterol 2000;95(4):859–60.

[10] Munyaradzi OM. Carcinoma of the stomach and large bowel – a comparative study of the incidence in blacks and whites in Zimbabwe/Rhodesia. S Afr J Surg 1979;17(4):179–87.

[11] Angelo N, Dreyer L. Colorectal carcinoma – a new threat to black patients? A retrospective analysis of colorectal carcinoma received by the Institute for Pathology, University of Pretoria. S Afr Med J 2001;91(8):689–93.

[12] Bonnet GC, Grobler SP. Disease profile of colon and rectum carcinoma at Pelonomi Hospital, Bloemfontein. S Afr J Surg 1997;35(2):63–5 [discussion 5-7].

[13] Boytchev H, Marcovic S, Oettle GJ. The characteristics of large bowel cancer in the low-risk black population of the Witwatersrand. J R Coll Surg Edinb 1999;44(6):366–70.

[14] Degiannis E, Sliwa K, Levy R, Hale MJ, Saadia R. Clinicopathological trends in colorectal carcinoma in a Black South African population. Trop Gastroenterol 1995;16(4):55–61.

[15] Ntombela XH, Zulu B, Masenya E, Sartorius B, Madiba TE. Is the clinicopathological pattern of colorectal carcinoma similar in the state and private health care systems of South Africa? Analysis of a Durban colorectal cancer database. Trop Doct 2017;47(4):360–4.

[16] van't Hof A, Gilissen K, Cohen RJ, et al. Colonic cell proliferation in two different ethnic groups with contrasting incidence of colon cancer: is there a difference in carcinogenesis? Gut 1995;36(5):691–5.

[17] Pucciarelli S, Agostini M, Viel A, et al. Early-age-at-onset colorectal cancer and microsatellite instability as markers of hereditary nonpolyposis colorectal cancer. Dis Colon Rectum 2003;46(3):305–12.

[18] Burkitt DP, Walker AR, Painter NS. Effect of dietary fibre on stools and the transit-times, and its role in the causation of disease. Lancet 1972;2(7792):1408–12.

[19] Irabor DO. Colorectal carcinoma: why is there a lower incidence in Nigerians when compared to Caucasians? J Cancer Epid 2011;2011:675154.

[20] Kampman E, Slattery ML, Bigler J, et al. Meat consumption, genetic susceptibility, and colon cancer risk: a United States multicenter case-control study. Cancer Epidemiol Biomark Prev 1999;8(1):15–24.

[21] Irabor D, Adedeji O. Colorectal cancer in Nigeria: 40 years on. A review. Eur J Cancer Care 2009;18(2):110–5.

[22] Segal I, Cooke S, Hamilton D, Tim LO. Polyps and colorectal cancer in South African Blacks. Gut 1981;22(8):653–7.

[23] de Mestier L, Manceau G, Neuzillet C, et al. Primary tumor resection in colorectal cancer with unresectable synchronous metastases: a review. World J Gastrointest Oncol 2014;6(6):156.

[24] Buie WD, Attard JA. Follow-up recommendations for colon cancer. Clin Colon Rectal Surg 2005;18(3):232–43.

[25] Davila RE, Rajan E, Baron TH. ASGE guideline: colorectal cancer screening and surveillance. Gastrointest Endosc 2006;63(4):546–57.

[26] Cohen AM, Kelsen D, Saltz L, et al. Adjuvant therapy for colorectal cancer. Curr Probl Surg 1997;34(8):601–76.

[27] Levin B, Lieberman DA, McFarland B, et al. Screening and surveillance for the early detection of colorectal cancer and adenomatous polyps, 2008: a joint guideline from the American Cancer Society, the US Multi-Society Task Force on Colorectal Cancer, and the American College of Radiology. CA Cancer J Clin 2008;58(3):130–60.

[28] Nelson RL, Dollear T, Freels S, Persky V. The relation of age, race, and gender to the subsite location of colorectal carcinoma. Cancer 1997;80(2):193–7.

[29] Mekenkamp L, Koopman M, Teerenstra S, et al. Clinicopathological features and outcome in advanced colorectal cancer patients with synchronous vs metachronous metastases. Br J Cancer 2010;103(2):159–64.

CHAPTER

15 ASPECTS OF VIRAL HEPATITIS IN SUB-SAHARAN AFRICA IN THE 21ST CENTURY

Mark W. Sonderup*, **C. Wendy Spearman**†

Division Hepatology and Department of Medicine, University of Cape Town, Cape Town, South Africa Division of Hepatology, Faculty of Health Sciences, Department of Medicine, University of Cape Town, Cape Town, South Africa*†

INTRODUCTION

The burden of liver disease in sub-Saharan Africa is mostly underestimated with the burden compounded by the lack of screening and access to care and treatment as well as inadequate disease surveillance, human and financial resources [1]. Data suggest that approximately 75 million have chronic HBV and 10.1 million have chronic HCV infection in sub-Saharan Africa [2, 3].

Viral hepatitis, in particular chronic infection, is a silent and neglected cause of morbidity and mortality aggravated by inactivity at a global level with respect to policy implementation. There is however new impetus with the advent of effective prevention through vaccination and the development of highly effective therapies as well as a WHO endorsed plan to eliminate viral hepatitis [4, 5]. This chapter looks at the unique epidemiological factors and treatment of viral hepatitis on the African continent and how aspects of this will likely change in the 21st century.

HEPATITIS A

Sub-Saharan Africa is a hyperendemic area for hepatitis A virus (HAV) with most primary infections occurring before the age of 10 years [6]. HAV is a picornavirus, is nonenveloped, and contains a single-stranded RNA. There is only one serotype, but multiple genotypes with human genotypes numbered I-III with 6 subtypes (IA, IB, IIA, IIB, IIIA, and IIIB) exist. HAV is spread via the fecal/oral route from an infected person. The incubation period of hepatitis A is usually 14–28 days, while symptoms can range from being subclinical or mild to severe and can include malaise, poor appetite, diarrhea, nausea, mild abdominal discomfort, and jaundice. Invariably, adults are more inclined to present with icteric features and symptoms as opposed to children who mostly experience subclinical infection.

Several hepatitis A vaccines are available with similar efficacy in terms of protection with nearly 100% of people developing protective levels of antibodies within 1 month of a single dose of the vaccine. No vaccine is licensed for children younger than 1 year of age and hence vaccination is advised in children >1 year old.

Digestive Diseases in Sub-Saharan Africa. https://doi.org/10.1016/B978-0-12-815677-3.00015-3

In sub-Saharan Africa, given its endemicity, outbreaks of hepatitis A are uncommon and hence routine vaccination for HAV has not been recommended [5]. Rather, the World Health Organization has suggested that individual countries should assess their hepatitis A IgG seroprevalence status and consider inclusion of the vaccine into their own immunization schedules based on their data. This issue however should be re-evaluated in the African context given socioeconomic transitions.

The incidence of hepatitis A is strongly correlated with access to clean water and adequate sanitation and this factor has mostly contributed to the endemic nature of hepatitis A in sub-Saharan Africa. Environmental Stability was the 7th Millennium Development Goal to be achieved by 2015 and 64% of those in sub-Saharan Africa now have access to an improved drinking water source. However, significant numbers are still reliant on unimproved or surface water that is fecally contaminated. Even so, we may be seeing the first shifts in the epidemiology of water-borne diseases such as hepatitis A. Work from West Africa has demonstrated that prior to 1990, most preschool children had evidence of immunity in keeping with the known epidemiology of most infections occurring before age 5. However, seroprevalence data from urban children have indicated that the midpoint of population immunity has shifted to school-aged children suggesting a transition to lower levels of hepatitis A endemicity [7]. This most likely reflects access to improved water sources and sanitation. This epidemiological transition with changing socioeconomic circumstances as well as the high HIV burden warrants an urgent review of policy with respect to hepatitis A vaccination in immunization schedules in the near future.

HEPATITIS B

Hepatitis B is an important public health issue in sub-Saharan Africa with the HIV/AIDS pandemic, complicating the natural history hepatitis B. Singly, hepatitis B is responsible for 70% of the cases of hepatocellular carcinoma that occurs in Africa and of greater importance, is wholly preventable through vaccination. The spectrum of disease and natural history of chronic hepatitis B virus (HBV) infection is diverse, ranging from a low viremic inactive "carrier," latency or immune control state to progressive chronic hepatitis with the potential ensuing complications of cirrhosis, liver failure, and hepatocellular carcinoma [8]. Over the past decade, our understanding of the natural history of HBV has significantly improved mirrored by significant therapeutic advances.

Hepatitis B is an enveloped partially double-stranded DNA virus belonging to the *Hepadnaviridae* family. It is 100 times more infectious than HIV and is transmissible via perinatal, percutaneous, or sexual means. Close person-to-person contact is an important form of transmission, most notably among children in highly endemic areas such as in sub-Saharan Africa. The risk of chronicity is age dependent with 90% of neonates born to HBeAg-positive mothers, 20%–50% of children under the age of 5 years, and approximately 5% of adults developing chronic hepatitis B infection following acute exposure to HBV [9].

Following acute exposure, the viral DNA enters the hepatocyte and translocates to the nucleus. The partially doubled-stranded DNA is repaired to form a circular extrachromosomal structure called covalently closed circular dsDNA (*cccDNA*), which is the transcriptional template for the viral messenger RNAs. The virus replicates its DNA genome by reverse transcription of an RNA intermediate via viral reverse transcriptase, which occurs within the cytoplasm. This enzyme is the major target of nucleos(t)ide analogues used in the management of chronic hepatitis B virus infection. The *cccDNA* is continually replenished and intercalated into the hepatocyte genome. Hence, even if the individual clears

HBsAg, the hepatocyte still harbors *cccDNA* that may allow for full viral reactivation in the presence of immunosuppression. This has implications for patients in sub-Saharan Africa where HBV is endemic and up to 90% of people in rural areas who are HBsAg negative may demonstrate serological evidence of past exposure, viz. anti-HB core IgG positive. Firstly, the presence of occult HBV infection, i.e., detectable HBV DNA at a very low level (usually <200 IU/mL) in the blood of those HBsAg negative and IgG anti-HB core positive, is higher in HIV-infected people [10]. With the high HIV prevalence sub-Saharan Africa, the risk for HBV reactivation and hepatocellular carcinoma is enhanced. More importantly, the risk of perinatal HBV transmission is increased. Such issues, together with higher HBV DNA viral levels in HIV/HBV coinfected people, suggests a compelling case for birth dose vaccination for hepatitis B in sub-Saharan Africa [11]. Only 23% of countries in the WHO Africa region have a birth dose vaccine as part of their expanded immunization schedule, most countries opting for 3 doses commencing 4 to 6 weeks after delivery. The issue of a wide scale birth dose-based HBV vaccine schedule in Africa is a serious matter that warrants urgent prioritization with support and funding [12]. Countries such as China have introduced such changes to their HBV vaccination schedule with success being demonstrated.

The natural history of chronic hepatitis B virus infection is characterized by several phases. These are determined after clinical, serological, and biochemical evaluation of a given patient. Phases include an immune-tolerant (HBeAg positive, high HBV DNA levels, normal transaminases), immune clearance or immune active (HBeAg positive, fluctuating HBV DNA, fluctuating transaminases), immune control or latent phase (HBeAg negative, low/undetectable HBV DNA levels, and normal transminases), and an immune-reactive HBeAg-negative phase (HBeAg negative, fluctuating/rising HBV DNA, fluctuating/rising transaminases). The immune-reactive phase poses a significant risk of progressive fibrosis and progression to cirrhosis with an associated even higher risk of hepatocellular carcinoma. Understanding the natural history is key to determining which patients warrant therapy or not.

To evaluate patients appropriately requires access to basic virological, serological, biochemical, and radiological (e.g., ultrasound) services. The gold standard for evaluating the liver is the liver biopsy that allows for assessing the degree of necroinflammation and fibrosis. However, in many parts of the continent, even these basic services are unavailable and liver biopsy with proper histopathological evaluation is unavailable. Other noninvasive tests of fibrosis such as Transient Elastography (Fibroscan^{R}), APRI or *Fib-4*, are useful alternatives. Transient Elastrography requires the availability of expensive equipment and laboratory testing is still required to perform other tests. A significant challenge is to develop cheap yet effective and good-quality diagnostic tests that can be used to evaluate large populations to determine the need for therapy. However, before this can even be achieved, hepatitis B needs to be regarded as a priority in sub-Saharan Africa. Seroprevalence data are very limited and no policy exists where testing for HBsAg is performed as part of routine ante-natal care, for example. Until screening and identification of those chronically infected is prioritized, little is likely to change.

In an attempt to start addressing those who need therapy and provide a framework in which this can happen in resource-constrained areas, the WHO launched their first set of guidelines in March 2015, aimed in particular at sub-Saharan African countries. The approach is the first endeavor to use noninvasive tests, age, ALT, and viral load as the only parameters in deciding who should access therapy now or wait. This approach avoids the need for checking HBeAg and compensates for a potential lack of availability of the histological assessment of a given patient by means of a liver biopsy.

The WHO recommendations suggest that all HBsAg-positive adults, adolescents, or children with evidence of compensated or decompensated cirrhosis regardless of the ALT, HBeAg status, or HBV viral load, must be offered treatment. Furthermore, treatment is indicated in adults without cirrhosis if they are older than 30 years with persistently elevated ALT levels and evidence of significant HBV replication (HBV viral load of >20,000 IU/mL). HBeAg is not required and HBV viral load is not a pre-requisite either. Of notable importance to sub-Saharan Africa, the suggested age of older than 30 years is also not prescriptive given that patients with advanced disease in the African context are often much younger.

Traditionally, therapy options in patients with chronic hepatitis B virus infection center on interferon-based immunomodulatory therapy or antiviral therapy with nucleos(t)ide analogues. Interferon is a very attractive option, notably for younger patients who are HBeAg positive with elevated ALT, moderately elevated HBV viral loads, and increased histological necroinflammatory activity. Interferon is also a finite course of therapy. However, several critical factors make this option unrealistic in sub-Saharan Africa. Firstly, the need for extensive diagnostics and workup in deciding who the ideal candidate for interferon is. Secondly, the need for close monitoring and follow up while on therapy. Thirdly and likely the most compelling reason against using interferon is that young Africans who are HBeAg positive, often do not meet the biochemical and histological ideals for interferon-based therapy given their lengthy period of immune tolerance in chronic HBV infection. This makes the use of antiviral therapy a more compelling if not a more sensible option in this setting [12].

Antiviral nucleos(t)ide analogues are potent inhibitors of HBV DNA replication and over the past decade, five nucleos(t)ide analogues have been globally licensed. These include lamivudine, telbivudine, adefovir, tenofovir, and entecavir. In sub-Saharan Africa, lamivudine and tenofovir is the most widely available given its dual role in the management of HIV/AIDS. Entecavir is generally not widely available despite it coming off patent >5 years ago and affordable generic options are not available. This is an ongoing challenge in sub-Saharan Africa and requires intervention at a global level. The best choice is either tenofovir or entecavir as they have a high genetic barrier to resistance. Although lamivudine is a very effective drug, its Achilles heel is the development of resistance that can reach 70%–90% after 5 years of therapy.

With tenofovir (as tenofovir disiproxil fumarate; TDF), the recommended dosage for adults with normal renal function (creatinine clearance >50 mL/min) is 300 mg per day. Dosage reduction is necessary in patients with impaired renal function and given its potential, albeit low, for renal toxicity, patients with known renal dysfunction should probably not use tenofovir. With entecavir, the recommended dose for adults with normal renal function (creatinine clearance >50 mL/min) is 0.5 mg daily if lamivudine naïve and 1 mg daily if previously exposed to lamivudine or if lamivudine-refractory or resistant. Dosage reduction is also necessary in patients with impaired renal function; however, entecavir has a negligible risk of renal toxicity and is the preferred antiviral in those with impaired renal function. A new formulation of tenofovir, tenofovir alafenamide or TAF, is dosed at 25 mg daily. Cohort data, now reporting at 152 weeks, demonstrate TAF to be virologically noninferior to TDF, however with a far superior renal and bone density toxicity profile [13]. TAF will replace TDF in due course as 1st-line therapy for those with chronic HBV infection.

Ideally, in those patients where HBeAg status is known, is positive and anti-HBe seroconversion occurs, treatment should be continued for at least 1 year prior to considering withdrawal. Careful follow-up after the cessation of successful treatment is important as up to 20% may relapse and serorevert to HBeAg positivity again. Patients with HBeAg-negative chronic hepatitis B and those with cirrhosis require lifelong treatment with antivirals. Given the need for ongoing follow up in those who are HBeAg positive, it is perhaps more sensible to just continue therapy indefinitely.

In patients who are HIV/HBV coinfected, a combination of lamivudine (or emtricitabine) and tenofovir forms the backbone of the antiretroviral therapy (ART) regimen. This should be continued irrespective if the ART regimen is amended in a given patient due to HIV resistance. This way, HBV replication remains suppressed. An anomalous paradox that exists in many African countries, whose HIV/AIDS programs are externally funded by international donors, is that drugs such as lamivudine and tenofovir are available to HIV/HBV coinfected patients, but unavailable to those with HBV only infection should they warrant therapy. This situation creates "exceptionalism" for those coinfected and is neither morally nor ethically acceptable for patients with HBV who need access.

HEPATITIS C

Worldwide approximately 71 million people are actively viremic with the hepatitis C virus (HCV) creating a significant contribution to the global burden of disease for a virus first sequenced a quarter century ago in 1989 [3]. Given no preventive or protective vaccine yet for HCV, measures to control the epidemic are focused on the known risks of HCV acquisition. Improved blood transfusion donor screening has all but eradicated this as a means of HCV transmission and needle exchange programs and opiate substitution therapy has reduced risk from people who inject drugs. The epidemiology of hepatitis C in sub-Saharan Africa is incompletely understood with accurate epidemiological information being very limited [14]. Furthermore, few data on the burden of coinfection with HIV exist. Significant regional variation has been observed with highest burden in Nigeria and parts of Central and Eastern Africa. Modeling studies suggest viremic rates varying between 1.3% (Nigeria) to 7% (Gabon).

Even so, more people in Africa are infected with hepatitis C than have HIV, yet the public health burden of hepatitis C has only now started to be recognized. For hepatitis C, the 2 major challenges for the 21st century are understanding the epidemiology in sub-Saharan Africa so that targeted screening can occur to identify those infected as well as getting affordable access to the new direct acting antiviral (DAA) drugs that offer almost universal cure using short duration regimens. The traditional modes of HCV transmission typically are prior (usually pre-1992) blood and blood product exposure or injecting drug use. Other less dominant modes have included unsafe medical procedures or equipment use, parenteral injuries, e.g., needle stick injuries in healthcare workers, sexual transmission, body piercings, and mother-to-child transmission. A vexing question remains: what factors account for the dominant modes of HCV transmission in sub-Saharan Africa? Approximately 8% of the global injecting drug users reside in sub-Saharan Africa and likely account for driving ongoing viral transmission but contribute little to the existing older aged hepatitis C population [15]. Blood donor screening and blood transfusion services in Africa are challenged and previously only 19% of blood was screened for HCV given the prohibitive costs of laboratory testing. The high HCV prevalence of 17% reported in patients with sickle cell patients who have received multiple blood transfusions suggests the significance of blood transfusions as a major risk for HCV transmission. With funding from PEPFAR, the Global Fund, and WHO, blood safety programs in 38 sub-Saharan African countries were introduced. This had the effect that between the time periods 2000/2004 and 2010/2011, the median annual number of units of blood donated per country increased, and the number of countries screening at least 95% of blood donations for HBV and HCV increased. Overall, HBsAg screening increased from 76% to 94% with HCV antibody screening increasing from 34 to 86%. The median percentage of HBV and HCV Ab marker-reactive blood donations decreased from 7.1% to 4.4% and 1.4% to 0.9%, respectively.

The WHO estimates that in sub-Saharan Africa approximately 18% of injections given are with reused syringes or unsterilized needles. The role that traditional practices, e.g., scarification rituals or circumcision, play is very difficult to quantify, suffice to say that it likely contributes to risk of HCV transmission. Vertical transmission is thought to be low, but possibly has increased significantly in the setting of coinfection with HIV. Puzzlingly, when one broadly views a map of Africa, there is a seemingly inverse relationship between HIV and HCV prevalence with HIV prevalence being high in Southern Africa and HCV prevalence rising as one moves more north in Africa [16]. The reasons for this are not entirely clear.

Phylogenetic analysis has determined 11 HCV genotypes, 1–11 with many subtypes a, b, c, etc. Only genotypes 1–6 are clinically relevant with genotypes 1–3 widely distributed globally and genotypes 1a and 1b accounting for 60% of infections worldwide. Genotype 4 dominates in the Middle East, Egypt, and Central Africa with genotype 5 almost exclusively found in South Africa [17].

The primary goal of therapy in patients with acute or chronic hepatitis C infection is to cure the infection. Irrespective of the means by which cure is achieved, it is sustainable over the medium to long term in >99% of individuals. Up to 2013, the standard of care was based on the use of pegylated interferon and ribavirin, used for between 24 and 48 weeks. Viral undetectability 24 weeks after the end of treatment heralded the attainment of a sustained virological response (SVR), which equated to a cure. However, treatment with weekly pegylated interferon and daily ribavirin was complicated by many adverse effects and the need for close follow up and monitoring. Furthermore, those with decompensated liver disease were precluded from using interferon-based therapy. While response rates in achieving SVR were promising, almost half of those being treated did not respond to or tolerate therapy. In Africa, access to therapy was further complicated by the relatively high cost of the therapy and need for extensive diagnostic workup in determining genotype, viral load, degree of liver disease, and laboratory follow up while on therapy.

The development of effective new drug targets for therapy was hampered by the lack of an animal model for HCV. Through cell culture-based techniques and models, the complete life cycle of HCV was elucidated in 2005. This spurred a revolution with 100 new drug targets being evaluated by 2010. Currently, the 3 major targets of therapy are the HCV NS3/4A protease (the NS3/4a protease inhibitors), the NS5a replication complex (NS5a inhibitors), and the NS5B RNA polymerase (nucleoside and non-nucleoside inhibitors). The first and second generations of NS3/4A protease inhibitors (telaprevir, boceprevir, simeprevir, paritaprevir) are mostly genotype specific and are active against genotype 1 and some, in addition, against 4. Newer NS3/4A protease inhibitors (grazoprevir, voxilaprevir) are pangenotypic. NS5a inhibitors (ledipasvir, daclatasvir, ombitasvir, velpatasvir, ruzasvir, pibrentasvir) have extended genotype activity, with the polymerase inhibitors mostly pangenotypic (sofosbuvir, dasabuvir, uprifosbuvir). All the regimens employ a combination of the drugs either in dual or triple combinations for durations of therapy currently ranging from 8 to 24 weeks, depending on the presence or absence of cirrhosis, treatment history, and ability to tolerate add-on ribavirin if required. Irrespective of the combination used, data suggest SVR rates of >90% for approved regimens [18].

No DAA regimens have been approved as yet for children with chronic hepatitis C. A phase 2, multicenter, open-label study evaluated the efficacy and safety of a 12-week course of ledipasvir-sofosbuvir in 100 adolescents aged 12 to 17 years (median age: 15 years) with chronic HCV genotype 1 infection. 98% achieved an SVR12 with no serious adverse events [19, 20].

Combinations of universally effective pangenotypic regimens that eliminate the need for complex diagnostics would revolutionize the management of chronic HCV infection in sub-Saharan Africa [14].

The need to determine genotype and pretreatment viral load would be eliminated allowing only for an effective test to determine the presence of active HCV infection or not. To do this, three things need to happen with some degree of urgency: (1) simple screening tests need to be validated in large resource-constrained populations; (2) at-risk populations need to be identified and screened; and (3) affordable and quality generic drugs need to be made available to treat those needing therapy. For a chronic viral infection that was only discovered in the latter part of the last century, the speed at which curative therapy has been achieved is astonishing. In this century, the challenge is translating that achievement to those who need it.

HEPATITIS D

HDV is a unique RNA virus that is dependent on HBV for survival. The HDV virion is a 36-nm particle and encloses a single-stranded circular RNA genome of approximately 1700 nucleotides within the HBsAg for replication and transmission. HDV is a defective virus, it does not encode its own replicase and is totally dependent on the host replicative machinery for its synthesis. Thus, there are no viral replicative enzymes for drugs to target. HDV is dependent on HBV for infection with HBV providing HBsAg to coat its virion. Infectivity depends on a receptor-binding domain in the N-terminal region of the pre-S1 of the large HBsAg, which specifically interacts with sodium taurocholate cotransporting polypeptide, a multiple transmembrane transporter predominantly expressed in the liver and does not depend on HBV DNA levels.

The HDV genome contains a ribozyme and encodes for 2 proteins, S-HDAg and L-HDAg The ribozyme, an RNA segment of approximately 85 nucleotides, which self-cleave and self-ligate without the need of enzymatic functions, is critical to the life cycle of the virus. S-HD Ag supports replication, and L-HD Ag promotes virion packaging while inhibiting HDV replication. S-HDAg and L-HDAg inhibit HBV replication by repressing hepatitis B virus enhancers and activating the alpha/beta interferon-inducible MxA gene. The pathogenic mechanism of HDV-induced liver damage is thought to be immune mediated and not due to a direct cytopathic effect of HDV.

Worldwide, approximately 15–30 million people are infected with hepatitis D. Although vaccination and prevention of HBV has reduced HDV seroprevalence in many populations, HDV remains a significant public health problem in the developed world.

In Africa, where HBV is endemic, documented HDV seroprevalence rates vary geographically from low rates in countries south of the Equator: 0%–0.6% in South Africa, 0% in Mozambique, 0.6% in Kenya (south of Mount Kenya) to high rates north of the equator: 31% in Kenya (north of Mount Kenya), 7%–44% in Tunisia, 25%–28% in Sudan, 17%–56% in Somalia, 3% in Senegal, 6%–12% in Nigeria, 15%–33% in Mauritania, 9%–67% in Gabon, 5%–58% in Egypt, 2%–22% in Djibouti, and 6%–27% in Cameroon. Epidemic outbreaks have been reported in Gabon, Cameroon, Nigeria, and Mauritania [21].

8 HDV genotypes have been identified and are associated with variable clinical courses: Genotype 1 is found worldwide (range of liver disease severity), Genotype 2 occurs in Japan, Taiwan, Russia (milder disease); Genotype 3 occurs in the northern countries of South America (more severe disease including fulminant presentations) and Genotype 4 occurs in Japan and Taiwan (milder disease). Genotypes 5-8 have been identified in African patients and are associated with HBV genotypes A-E, but little is known about the clinical course of the liver disease.

HDV is transmitted via the parenteral route and similar to HBV, an important risk factor in sub-Saharan Africa is cohabitation with an HDV-infected HBsAg carrier in the setting of an overcrowded family household. Sexual transmission occurs and the risk of perinatal transmission is thought to be low. IUD is also an important risk factor for triple infection with HBV, HDV, and HIV.

Clinically, HDV infection can present as a coinfection, superinfection, or chronic infection with HBV. If acute HBV/HDV coinfection is acquired in adulthood, >95% individuals will clear both HBV and HDV, although there is a greater risk of fulminant hepatitis than with acute Hepatitis B [22].

Superinfection with HDV can present as an acute hepatitis in a previously asymptomatic HBsAg carrier or result in further clinical deterioration in individuals with established HBV disease. Hepatitis D becomes chronic in 70%–90% individuals with superinfection and there is more rapid progression to cirrhosis and decompensation, especially in IDU where end-stage liver disease can occur in <2 years. There is also an increased risk of hepatocellular carcinoma. Liver transplantation for HBV/HDV cirrhosis has a better prognosis than HBV cirrhosis. Prophylaxis with hepatitis B immune globulin and antiviral HBV therapy successfully protects against HDV and HBV reinfection as HBV replication is usually markedly diminished.

Coinfection can be distinguished from superinfection and chronic infection by serological and virological testing. In superinfection, markers of acute hepatitis D (HDV RNA, IgM, and IgG anti-HD) are present in addition to the serological and virological markers of chronic hepatitis B that will vary according to the phase of chronic hepatitis B infection. In HBV/HDV coinfection, markers of both acute hepatitis B (HBsAg, HBeAg, IgM anti-core Ab, and HBV DNA) and acute hepatitis D are detectable. The markers of HDV infection decrease rapidly and disappear after the clearance of HBsAg. Anti-HD disappears within a few months to a few years in most patients, but may persist.

Once chronic hepatitis D has developed, HBV replication is usually suppressed (low or undetectable HBV DNA) and HBeAg is negative. HDV markers of chronic infection include anti-HD, IgM anti-HD, and HDV RNA. In HBV/HDV/HCV triple infection, HDV is usually the dominant virus, inhibiting the replication of HBV and HCV.

There is no vaccine against HDV, but HBV vaccination is effective prophylaxis against HDV. The current recommended treatment of compensated chronic hepatitis D disease according to expert opinion is peginterferon alfa given weekly for 48 weeks, leading to HDV RNA clearance in 17%–47%, but relapse is common [23]. Low baseline HBsAg and HDV RNA titers may predict a positive response to therapy. Control of HDV infection may be associated with HBV reactivation requiring antiviral therapy.

The HIDIT-1 (adefovir) and HIDIT-11 (tenofovir) trials showed that the addition of antivirals did not improve the efficacy of pegylated interferon therapy. Novel therapies in clinical development are directed at targeting: (1) Process of prenylation of the large HD antigen. *Lonafarnib,* a prenylation inhibitor, prevents the assembly of the hepatitis D virion by blocking the combination of HBsAg with the large hepatitis D antigen; (2) Attachment of HBsAg to the hepatocytes by the sodium-taurocholate cotransporting polypeptide. *Myrcludex B,* a synthetic N-acylated pre-S1 lipopeptide, which blocks in vivo and in vitro the receptor function of sodium-taurocholate cotransporting polypeptide, thereby preventing the entry and spreading of HDV to uninfected hepatocytes; (3) Synthesis of the HBsAg in the liver with *REP-2139*, a nucleic acid polymer, that blocks the assembly of HBV subviral particles leading to diminished HDV virion morphogenesis.

Studies have shown that *Lonafarnib* and *Myrcludex B* reduced serum HDV-RNA; but neither diminished serum HBsAg. *REP-2139* diminished both HDV-RNA and HBsAg in serum [24]. These new drugs may provide additional efficacy in combination with Peg-Interferon.

HEPATITIS E

In 1955, an outbreak of clinical viral hepatitis in New Delhi occurred during the Monsoon season affecting almost 30,000 people. No cause could be identified despite stool samples being subjected to electron microscopy at the NIH. Hepatitis E, an RNA virus, was only formally identified in 1983 after an outbreak in Afghanistan where a Russian doctor made a yoghurt shake with a patient's stool. Once he became symptomatic, he gathered stool samples and the virus was isolated and identified.

Hepatitis E infects about 20 million people annually with a 0.2%–4% mortality rate causing approximately 70,000 deaths. It is predominantly encountered in Southern Asia and parts of West, North-East, and Southern Africa [25]. There is one serotype of hepatitis E and 4 genotypes, 1–4. Genotypes 1 and 2 are responsible for epidemic outbreaks and are mostly associated with contaminated water although person-to-person transmission is described. Infection is restricted to humans with genotype 1 predominating in Asia and genotype 2 in Africa and Mexico. Risk of symptomatic infection is greatest in pregnancy and men although most of those infected are asymptomatic. Genotypes 3 and 4 are primarily a zoonosis with mild disease in humans, the overwhelming majority asymptomatic. It is associated with eating pork, deer, and mussels. Risk for symptomatic disease is seen in older men and hepatitis B infected individuals. Genotype 3 has been described in pigs in Cameroon and the Central African Republic [26]. In addition, a high seroprevalence of hepatitis E was noted in pig handlers in Ghana. In South Africa, four hepatitis E cases have been identified. In Africa, most outbreaks are sporadic with seroprevalence data from Zambia suggesting an overall seroprevalence of 42%; 28% in HIV-negative, and 71% in HIV-positive adults. A recent study in the Western Cape in South Africa in both adults and children found an overall seroprevalence of approximately 30% [27].

Interestingly, genotype 3 has been associated with chronic infection with viral RNA persisting beyond 6 months. This has been described almost exclusively in immunecompromised people such as those post solid organ transplant with about 10% of those affected developing progressive liver disease with advanced fibrosis or cirrhosis.

Given its association with contaminated water, the issues around access to safe and clean water sources for preventing hepatitis A are synonymous for hepatitis E. Given genotype 3 and its association with raw or undercooked meat, mostly pork, basic hygiene principles of avoiding undercooked meats is critical to prevention. Two vaccines have been through phase I and II studies but only one commercial vaccine is available, although it is only available in China. A randomized study of over 100,000 adults in China demonstrated 100% antibodies to hepatitis E after 3 doses of vaccine [28]. As yet the vaccine is not registered elsewhere but use would be advised in endemic areas, in the immunosuppressed, those with chronic liver disease and possibly food handlers. Even so, the clinical positioning of the vaccine is yet to be determined.

REFERENCES

[1] Spearman CW, Sonderup MW. Health disparities in liver disease in sub-Saharan Africa. Liver Int 2015;35(9):2063–71.

[2] Schweitzer A, et al. Estimations of worldwide prevalence of chronic hepatitis B virus infection: a systematic review of data published between 1965 and 2013. Lancet 2015;386(10003):1546–55.

[3] Polaris Observatory HCV Collaborators. Global prevalence and genotype distribution of hepatitis C virus infection in 2015: a modelling study. Lancet Gastroenterol Hepatol 2017;2(3):161–76.

[4] WHO, http://apps.who.int/iris/bitstream/handle/10665/206453/WHO_HIV_2016.04_eng.pdf;jsessionid=5B91B2C7A523D1FC08170714D58608EB?sequence=1; 2016. (Accessed April 26, 2018).
[5] Mohd Hanafiah K, Jacobsen KH, Wiersma ST. Challenges to mapping the health risk of hepatitis A virus infection. Int J Health Geogr 2011;10:57.
[6] Jacobsen KH, Wiersma ST. Hepatitis A virus seroprevalence by age and world region, 1990 and 2005. Vaccine 2010;28(41):6653–7.
[7] Jacobsen KH. Hepatitis a virus in West Africa: is an epidemiological transition beginning? Niger Med J 2014;55(4):279–84.
[8] Spearman CW, et al. South African guideline for the management of chronic hepatitis B: 2013. S Afr Med J 2013;103(5 Pt 2):337–49.
[9] Lin X, et al. Chronic hepatitis B virus infection in the Asia-Pacific region and Africa: review of disease progression. J Gastroenterol Hepatol 2005;20(6):833–43.
[10] Hoffmann CJ, Thio CL. Clinical implications of HIV and hepatitis B co-infection in Asia and Africa. Lancet Infect Dis 2007;7(6):402–9.
[11] Spearman CW, Sonderup MW. Preventing hepatitis B and hepatocellular carcinoma in South Africa: the case for a birth-dose vaccine. S Afr Med J 2014;104(9):610–2.
[12] Spearman CW, et al. Hepatitis B in sub-Saharan Africa: strategies to achieve the 2030 elimination targets. Lancet Gastroenterol Hepatol 2017;2(12):900–9.
[13] Wong GL, et al. Review article: long-term safety of oral anti-viral treatment for chronic hepatitis B. Aliment Pharmacol Ther 2018;47(6):730–7.
[14] Sonderup MW, et al. Hepatitis C in sub-Saharan Africa: the current status and recommendations for achieving elimination by 2030. Lancet Gastroenterol Hepatol 2017;2(12):910–9.
[15] Nelson PK, et al. Global epidemiology of hepatitis B and hepatitis C in people who inject drugs: results of systematic reviews. Lancet 2011;378(9791):571–83.
[16] Modi AA, Feld JJ. Viral hepatitis and HIV in Africa. AIDS Rev 2007;9(1):25–39.
[17] Prabdial-Sing N, et al. Hepatitis C genotype distribution in patient and blood donor samples in South Africa for the period 2008-2012. J Viral Hepat 2016;23(11):881–8.
[18] European Association for the Study of the Liver. EASL Recommendations on Treatment of Hepatitis C. J Hepatol 2018;2018 [in press], https://doi.org/10.1016/j.jhep.2018.03.026.
[19] Wirth S, et al. Sofosbuvir and ribavirin in adolescents 12-17 years old with hepatitis C virus genotype 2 or 3 infection. Hepatology 2017;66(4):1102–10.
[20] Balistreri WF, et al. The safety and effectiveness of ledipasvir-sofosbuvir in adolescents 12-17 years old with hepatitis C virus genotype 1 infection. Hepatology 2017;66(2):371–8.
[21] Stockdale AJ, et al. Prevalence of hepatitis D virus infection in sub-Saharan Africa: a systematic review and meta-analysis. Lancet Glob Health 2017;5(10):e992–e1003.
[22] Buti M, et al. Clinical outcome of acute and chronic hepatitis delta over time: a long-term follow-up study. J Viral Hepat 2011;18(6):434–42.
[23] Heller T, et al. Long-term therapy of chronic delta hepatitis with peginterferon alfa. Aliment Pharmacol Ther 2014;40(1):93–104.
[24] Rizzetto M. Investigational drugs in development for Hepatitis D. Expert Opin Investig Drugs 2017;26(9):999–1005.
[25] Kim JH, et al. A systematic review of the epidemiology of hepatitis E virus in Africa. BMC Infect Dis 2014;14:308.
[26] de Paula VS, et al. Hepatitis E virus genotype 3 strains in domestic pigs, Cameroon. Emerg Infect Dis 2013;19(4):666–8.
[27] Madden RG, et al. Hepatitis E virus: Western Cape, South Africa. World J Gastroenterol 2016;22(44):9853–9.
[28] Innis BL, Lynch JA. Immunization against Hepatitis E. Cold Spring Harb Perspect Med 2018.

CHAPTER

HEPATOCELLULAR CARCINOMA IN AFRICA

16

Mashiko Setshedi, Monique Andersson

Department of Medicine, University of Cape Town, Cape Town, South Africa

INTRODUCTION

The next few decades could herald dramatic reductions in the prevalence of HCC on the African continent. The increase in hepatitis B virus (HBV) vaccine coverage and the advent of safe, effective, and curative therapies for hepatitis C virus (HCV) will reduce the prevalence of these dominant risk factors. Significant challenges remain, however, not least the economic cost of the implementation of effective hepatitis control.

GLOBAL IMPORTANCE

Hepatocellular carcinoma (HCC) is the sixth most common cause of cancer and the fourth leading cause of cancer deaths worldwide [1]. In men, HCC is the second most common cause of death from cancer. An estimated 748,000 new cases of liver cancer occur annually and approximately 696,000 people die annually [2], resulting in an annual fatality ratio of 0.93, the highest of any malignancy. Sub-Saharan Africa (SSA) is one of the three geographical areas where HCC occurs very commonly [3].

EPIDEMIOLOGY IN SUB-SAHARAN AFRICA

To date data describing the occurrence of HCC in SSA have neither been complete, nor always accurate. As a result, the incidence of HCC is likely to be significantly understated, particularly as tissue diagnosis of HCC is rarely performed. Reports of a series of hospital-based patients have been and remain biased by local experience. Registry data are inaccurate and/or often not available. An age-standardized ratio of >15 per 100,000 of the population per year is regarded as a high incidence of HCC (Fig. 16.1). The age-standardized incidence rate in SSA is 41.2/100,000 persons/year: for South Africa it is 16.0 for males and 4.5 for females, for Zimbabwe 30.7 for men and 17.1 for females, and in Swaziland 22.0 in males and 5.2 in females [2]. By comparison, in 2010 the annual incidence of HCC was 6 per 100,000 in the USA (El-Serag 2014) [4].

The usual age distribution in high-income countries is late fifties to late seventies. In SSA, there is an obvious trend toward a younger age in Black males with a mean age of presentation between 33 years and 45 years, depending on whether one lives in a rural or urban area [5, 6]. Men have a higher

Digestive Diseases in Sub-Saharan Africa. https://doi.org/10.1016/B978-0-12-815677-3.00016-5

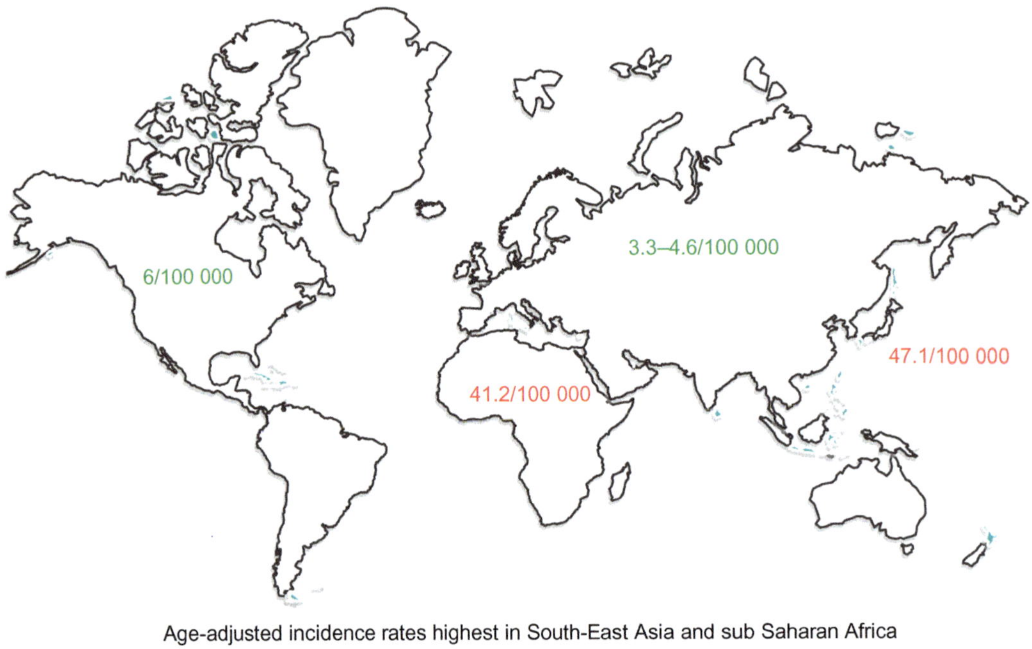

FIG. 16.1

Incidence of HCC in SSA compared to globally.

incidence of HCC in SSA than do women in all populations [1, 7]; the reasons are not well understood but probably reflect the higher rate of chronic HBV (approximately twice that in females) and higher alcohol intake. Dietary intake of carcinogens may be greater in men by virtue of the volume of food ingested and there may be differences in the efficiency with which ingested carcinogens are metabolized in men compared to women.

ETIOLOGICAL RISK FACTORS

The development of HCC is a multistep process with accumulations of genetic and epigenetic alterations in regulatory genes, leading to activation of oncogenes and inactivation or loss of tumor suppressor genes. Other pathways include aberrant signaling of pathways that regulate growth, differentiation, and angiogenesis, and the induction of a tolerogenic immune response [8].

VIRAL HEPATITIS

Viral hepatitis B and C infections together account for 75%–80% of cases of HCC globally, while HBV alone accounts for >50% of cases. The high prevalence of viral hepatitis mirrors those areas with a high prevalence of HCC [3]. Among individuals infected with HBV, the age at diagnosis of HCC is correlated with the age at acquisition and rate of viral replication.

HEPATITIS B

HBV belongs to the hepadnaviridae family, a family of enveloped DNA viruses with an incomplete double-stranded genome of 3.2 kb. The HBV genome contains four overlapping reading frames, which enable it to code for polymerase, surface, core, and x proteins. Previous work in Asian populations has shown that the risk of HCC is increased in those who have a high HBV viral load [9]. While there are little longitudinal data from SSA, there is evidence of significantly higher HBV viral load in Black African patients with HCC [10]. Specific mutations thought to play a role in the evolution of HCC include mutations in the basal core promoter, precore regions, and pre-S deletions. Little has been published on the mutations associated with HCC in SSA with genotype A1 (the most commonly found in SSA). The role of occult HBV infection in the carcinogenesis of HCC has been described, however is not well understood.

HBeAg status associated with a high HBV viral load carries an increased risk of HCC development. Chronic HBV carriers in SSA differ from those in Asia because the HBeAg seroconversion rate is much higher in Asia. This, however, may be changing in the context of HIV in that the prevalence of HBeAg has previously been thought to be around 5%, which is in comparison to more recent data showing HBeAg status that may be as high as 38% in patients with HBV infection [11].

Genotype A is the dominant genotype of HBV in South Africa, accounting for 70% of isolates. This genotype has characteristics that abrogate the formation of HBeAg. These mutations include T1762/A1764 or T1809/1812 or A1809/T1812 double mutation. Genotype A in the African patient is associated with an increased risk of HCC development. In South Africa, HBV genotype A increases the risk of HCC 4.5-fold compared to non-A genotypes [12].

HEPATITIS C

HCV is the most common cause of HCC in patients with cirrhosis suggesting that necroinflammation is important in the development of HCC. HCV is a small, positive sense single-stranded RNA virus from the Flaviviridae family. There are seven major genotypes that predict response to treatment and risk of HCC development.

AFLATOXIN

Aflatoxin (AFB1) is a metabolite produced by *Apergillus flavus* and *Aspergillus parasiticus;* it interacts with HBV to increase multiplicatively the risk of developing HCC [13]. It contaminates foodstuffs such as corn, rice, and peanuts. AFB1 covalently binds to guanine and cytosine residues of DNA and forms AFB1-DNA, RNA, and protein adducts which impair DNA, RNA, and protein synthesis, respectively. The evidence suggests that aflatoxin exposure and HCC are associated with mutations in p53 (frequently observed in human cancers), including a point mutation at the third position of codon 249 resulting in a G:C to T:A transversion. Mitotic recombinations and genetic instability are other mechanisms whereby aflatoxin may contribute to the development of HCC. HBV may directly or indirectly sensitize the hepatocyte to the effect of aflatoxin.

Aflatoxin-induced mutations may occur because of the increased hepatocyte proliferation that may promote clonal expansion of mutant cells. HBV x protein expressing cells are more prone to induction of mutations at codon 249 of the p53 gene. It has been proposed that HBV x protein could inhibit excision repair, leading to adduct persistence and mutation induction [14]. AFB1-albumin adducts in patients with HCC correlate significantly with the presence of DNA hypermethylation and mutations in p16 and p53 [15].

ALCOHOL

Excessive alcohol consumption is a major cause of HCC in high-income countries. The cause of HCC is thought to be due to chronic inflammation caused by alcohol-induced cirrhosis, as there is no convincing clear evidence that alcohol itself is hepatocarcinogenic. The risk of HCC development increases with increasing quantity over time. When the quantity exceeds 80 g for at least 5 years, the risk approaches 1% per year [16]. Approximately 15% of patients with alcoholic cirrhosis will develop HCC. Urban South African Blacks who habitually ingest >80 g of alcohol per day are at risk of developing HCC [17]. In rural Blacks, the cirrhosis that accompanies HCC is macronodular in type, which rarely has the features of alcohol misuse. In Gambia, only 13% of HCC patients report any alcohol intake [18]. An additive effect of HBV and excess alcohol consumption has been described. The odds ratio for tumor development in patients infected with HBV increases from 22.8 with alcohol intake of <60 g per day to 48.6 with consumption of greater than this amount [19].

CIRRHOSIS

The importance of cirrhosis in the development of HCC is well established [20]. The accelerated rate of cell division allows less time for mutated cellular DNA to be repaired and results in the accumulation over time of a number of mutations that are essential for tumor formation. In HBV carriers without cirrhosis, the risk ranges according to geographic region; from 0.4% to 0.6% per year in Asians to 0.2% or less in Caucasians. In those with cirrhosis, the risk is >2% per year, with a cumulative five-year incidence of 15%–20% [21]. In Black Africans, the role of cirrhosis in the evolution of HBV-related HCC is not well described.

IRON OVERLOAD

The detrimental effects of iron overload and HCC have been described for hereditary hemochromatosis and other iron overload syndromes. In an African setting, the iron overload secondary to drinking home-brewed traditional African beer is well reported. Portal fibrosis and occasionally cirrhosis may develop. Aside from cirrhosis, it is thought that the excess hepatic iron may be directly carcinogenic, with a reported relative risk of HCC of 10.6 (95% CI: 1.5–76.8) in a South African study after adjusting for major risk factors [22].

FAMILY HISTORY

Familial clustering of cases of HCC is well described. It remains an issue of debate as to whether this is due to common exposure to environmental factors or genetic factors that may predispose to the evolution of HCC. In a US study, individuals with a first-degree relative with HCC were four times more likely to develop HCC than those without a family history [23]. In a Taiwanese study of 553 HBV carriers with HCC and 4684 HBV carriers without HCC, the multivariate adjusted rate ratio was 2.41 (95% CI: 1.47–3.95). In Alaskan and Chinese studies, a single gene model has been suggested to explain the mode of inheritance in some families, although no genetic signature has been identified [24]. There are no published data on the role of family history in the evolution of HCC in individuals from SSA.

HIV AND ITS ASSOCIATION WITH HCC

In resource-rich settings liver disease has become, after opportunistic infections, the most common cause of death in patients treated with antiretroviral therapy (ART) [25]. Patients coinfected with HIV and HBV were shown to be 17 times more likely to die from liver-related diseases than those with HBV monoinfection [25]. Whether HIV increases the risk of HCC is less clear in resource poor settings. The largest study, a case referrant study, reported data on 1017 cancer cases, 60 had HCC and 12% (7 cases) were HIV infected. The OR for HCC in HIV-infected was 2.7 (95% CI: 1.1–7.7). The median age at presentation in HIV-infected patients was lower, 32 (IQR: 31–44) vs 49 years (IQR: 44–59) ($P<.001$) [26].

The theoretical mechanisms that might account for an increase in risk of HCC in the HIV-infected patient include the loss of immune control associated with HIV, resulting in higher HBV viral loads. HIV may directly facilitate malignant transformation in coinfected patients. Among transgenic mice, *Tat* protein expressed in the liver appears to enhance the malignant transforming potential of hepatitis viruses. *Tat*-binding protein interacts with HBx gene to regulate HBV transcription, and HBx protein induces HIV-1 replication [27].

CLINICAL PRESENTATION

SYMPTOMS

Early HCC is generally asymptomatic. Patients that are diagnosed early are typically patients with hepatitis B or C-related cirrhosis, who are under a screening program. Patients who are not being screened tend to present with late HCC that is not amenable to therapy. This is particularly true in sub-Saharan Africa, when patients tend to present with multicentric diffusely infiltrating HCC. When they present, however, the commonest and initial presentation is abdominal pain (50%–90%) [18]. The abdominal pain is typically in the right upper quadrant, but may be located in the epigastrium, or lower anterior chest. This may be confused with an amoebic liver abscess. The abdominal pain is caused by infiltration of the liver with tumor resulting in stretching of the capsule. Patients may report an enlarging abdomen with an obvious mass. The pain may be associated with abdominal distention due to ascites (although rare particularly in African patients) [28]. Other prominent symptoms include weight loss [18], malaise [28], and early satiety (in 25% of patients). Uncommon symptoms of HCC include jaundice, fever, bone pain, dyspnea, cough, and diarrhea [29].

SIGNS

The commonest physical sign is a firm or hard hepatomegaly [18, 28], which may occasionally be tender. A bruit may be heard over the liver, which is suggestive but not diagnostic of HCC. In late disease, patients may also present with tense ascites, with an enlarged ballotable liver [29]. An enlarged spleen

Table 16.1 Unusual Clinical Presentations of HCC

Unusual Clinical Features	Paraneoplastic Features
Acute abdomen due to rupture of bleeding tumor	Erythrocytosis
Budd-Chiari syndrome	Hypoglycemia
Superior vena cava syndrome	Hypercalcemia
Skin:	Hypercholesterolemia
Pityriasis rotunda	Thrombocytosis
Dermatomyositis	Watery diarrhea
Pemphigus folaceus	*Others:*
Seborrheic dermatitis	Carcinoid syndrome
Porphyria cutanea tarda	Osteoporosis
Pityriasis rotunda	Thyrotoxicosis

may be palpated, in a patient with portal hypertension [28]. In a patient with chronic liver disease with cirrhosis, the onset of decompensation with jaundice, variceal bleeding, ascites, or encephalopathy should prompt the possibility of HCC.

UNUSUAL PRESENTATIONS (TABLE 16.1)

Unusual clinical presentations of HCC include abdominal pain (due to rupture and bleeding of a necrotic tumor into the peritoneum) [28], which presents as an acute abdomen with hypotension and a drop in the hemoglobin. Management includes urgent embolization or surgery. HCC is a highly vascular tumor with the ability to infiltrate vascular structures, thus may present with a Budd Chiari syndrome, inferior [29] and superior vena caval obstruction, and more commonly, portal vein thrombosis, presenting with portal hypertension, in the latter. Skin manifestations of HCC are rare, but include pityriasis rotunda, dermatomyositis, pemphigus folaceus, seborrheic dermatitis (sign of Leser-Trelat), and porphyria cutanea tarda (in patients with hepatitis C infection) [30]. Pityriasis rotunda, also called circumscripta, is a round or oval scaly rash, typically seen in patients in SSA with HBV infection. The lesions are multiple, hyperpigmented, and are distributed on the trunk, lower abdomen, back, and buttocks. They are highly suggestive of HCC in this context, especially if seen in older patients. The commonest sites of extrahepatic spread are the lung, intra-abdominal lymph nodes, and bone, usually occurring in advanced disease.

PARANEOPLASTIC MANIFESTATIONS OF HCC

The commonest paraneoplastic manifestation is erythrocytosis, due to ectopic erythropoietin secretion by the tumor. Rare conditions include hypoglycemia (insulin-like growth factor II), hypercalcemia (parathyroid hormone-related protein), hypercholesterolemia, thrombocytosis, watery diarrhea (vasointesinal peptide, prostaglandin E2), and systemic hypertension [29]. Except for erythrocytosis, their presence is usually associated with a poorer prognosis [31]. Other rarer manifestations are listed in Table 16.2.

Table 16.2 Imaging for HCC in SSA

Modality	Pooled Sensitivities	Pooled Specificities	Advantages
Ultrasound	48% (95% CI: 34%–62%)	97% (95% CI: 95%–98%)	Useful for screening HCC - Safe - Cost effective
Multidetector CT	67.5% (95% CI: 55%–80%)	92.5% (95% CI: 89%–96%)	- Preferred method for diagnosis - Evaluates portal/hepatic veins - Determines location and number of lesions
Magnetic resonance imaging (MRI)	80.6% (95% CI: 70%–91%)	84.8% (95% CI: 77%–93%)	- May be slightly superior to CT

DIAGNOSIS

HCC is diagnosed noninvasively by typical features on imaging and where the criterion is not met, a liver biopsy. The use of alpha fetoprotein (AFP) for surveillance is now obsolete due to low sensitivity and specificity and has been dropped from international guidelines [32, 33]. In SSA, however, AFP is still used and is a useful and cost-effective means of diagnosing HCC. When AFP is raised, or shows an increasing trend in a patient, it is highly suggestive of HCC, although it is important to note that it may be normal in HCC. Due to increased surveillance, it is possible nowadays to diagnose HCC early in developed countries, however in sub-Saharan Africa, diagnosis is often made late, mainly due to the asymptomatic nature of HCC and late presentation of patients to hospital.

RADIOLOGICAL DIAGNOSIS

Different imaging modalities have their advantages and disadvantages (Table 16.2); in SSA specifically, ultrasound is still useful for screening for HCC. HCC (due to the blood supply of the liver and tumor) has typical features on imaging studies and differential appearance on imaging with contrast [32]. During the arterial phase, therefore, due to the fact that the tumor has an arterial blood supply relative to the mixed arterial and venous blood supply of the surrounding liver, HCC enhances with contrast. During the venous or delayed phase, on the other hand, due to the lack of venous supply in the tumor, compared to the surrounding liver tissue containing contrast in the portal veins, there is a so-called washout phase, which may persist in the delayed phase. To diagnose HCC, therefore, a 4-phase multidetector CT or dynamic contrast-enhanced MRI scan is required; it evaluates the uncontrasted, arterial, venous, and delayed phases. These typical features are highly specific for HCC and are sufficient to diagnose HCC in lesions >1–2 cm in diameter [34]. If these typical radiological features are absent in a lesion between 1 and 2 cm, a second and different contrasted radiological study should be performed [33]. If atypical features persist, a liver biopsy should be performed. In a lesion >2 cm with atypical radiological features, a biopsy should be performed immediately. Lesions <1 cm are monitored by 3 to 6 monthly ultrasound. If the lesion remains stable for 2 years, they may be surveyed at 6-month intervals. If the lesion changes in size or character, however, it should be investigated according to size. Differentiation of a dysplastic lesion from early HCC is not easy but critical because of the potential for cure in the latter, if recognized timeously.

TISSUE DIAGNOSIS

The diagnosis of small lesions (< than 2 cm) is complex because they may not have typical radiological features and histological diagnosis is not straightforward [35]. The histological diagnosis of HCC rests on the finding of stromal invasion, to differentiate them from dysplastic lesions, and this may not always be found. In this instance, a second biopsy may become necessary. These lesions can be further delineated by using biomarkers that are specific for liver tissue including CD34, CK7, glypican 3 (GPC3), and glutamine synthetase. GPC3 has a sensitivity of up to 72% and specificity >98% [36, 37] and is emerging as a promising marker for the histological diagnosis of HCC. Current recommendations are to use two biomarkers, which increase sensitivities and specificities [38]. To differentiate high-grade dysplastic from early HCC lesions, gene expression markers of biomarkers including survivin are recommended.

THE BARCELONA CLINIC LIVER CANCER STAGING SYSTEM (FIG. 16.2, [38A])

In order to prognosticate based on the stage of liver disease and HCC, staging systems have been developed. While several have been tested, only the Barcelona Clinic Liver Cancer (BCLC) staging system has been validated and endorsed by both the American Association for the Study of Liver Diseases (AASLD) [32] and European Association for the Study of the Liver (EASL) [33]. In addition, this system is able to assist in deciding on optimal therapy depending on stage of disease. Patients are divided into five groups according to tumor characteristics (size, number, tumor invasion, nodal involvement, and metastases), underlying liver function according to Child Pugh score, and general health status. BCLC stage 0 is a localized circumscribed tumor <2 cm in size without invasion or carcinoma in situ in patients with preserved liver function and good health status. These tumors have a good prognosis with a 5-year survival of up to 90% with resection or transplantation [39]. BCLC stage A patients have single tumors >2 cm or 3 nodules <3 cm in size, with Child Pugh A or B, but good health status. BLCL stage B represents an intermediate stage where patients are asymptomatic and present with multinodular noninvasive disease. The median survival without treatment is 16 months [40], but this can be extended with chemoembolization by 3 or 4 months. Advanced disease is categorized as BCLC stage C. These patients are symptomatic, have vascular invasion, extrahepatic spread, and poor health status. The prognosis in this group is very poor with a median survival of 6 months [41], with those with better liver function having a slightly better prognosis. Treatment with sorafenib has improved the survival period in these patients by 3 to 4 months. When patients are terminal, they have BCLC stage D. Here the patients have significant cancer-related symptoms with very poor health status. The median survival is only 3–4 months [42]. This group represents a large proportion of Black patients seen in SSA for whom the available therapies are not appropriate.

TREATMENT

Previously the treatment of HCC was discouraging because most patients presented with advanced HCC that was not amenable to treatment, and the treatment options themselves were limited and fraught with side effects. Nowadays there are a multitude of treatment options that are available, and

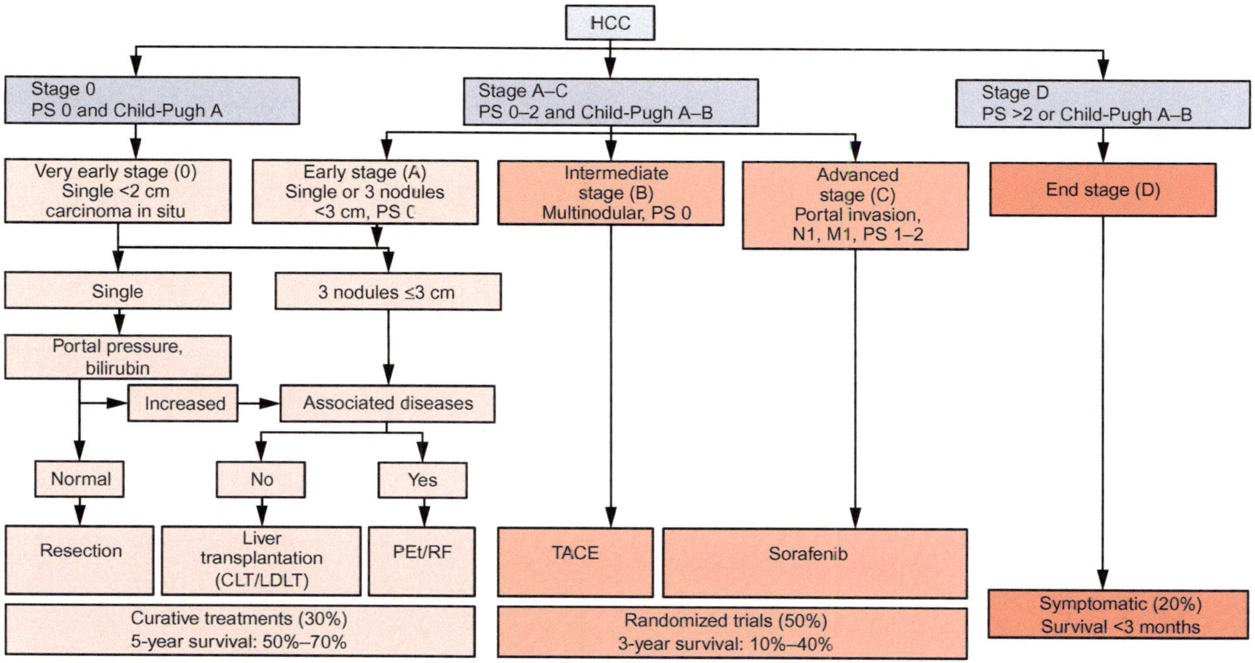

FIG. 16.2

Imaging for HCC in SSA.

early diagnosis and therefore curative treatment is possible. Due to the heterogeneity of HCC, patients must be managed by a multidisciplinary team consisting of hepatologists, hepatobiliary surgeons, radiologists, pathologists, and oncologists.

LIVER TRANSPLANTATION

Liver transplantation is the definitive curative treatment of choice; however, patients must be carefully selected. To this end, the Milan criteria were developed for selecting patients with HCC who had the highest chance of a good outcome with transplantation. The Milan criteria are: a solitary tumor not > than 5 cm, or less than three lesions, the largest being <3 cm in size [43]. The greatest risk of recurrence of tumor in the transplanted liver is proportionally related to the size and number of tumors; hence, the fewer and smaller in size the tumor, the higher the likelihood of a good post-transplant outcome. If these criteria are met, the 5-year survival is >70%. Given the restrictive nature of these criteria, there is a risk of denying patients with slightly bigger tumors that may still have a good outcome with transplantation, but due to the limited availability of organs, especially in Africa, this approach is reasonable, and the need for extended criteria for listing for HCC is reduced. In an attempt to prioritize patient's need for transplantation within a three-month period, the model for stage liver disease (MELD) score was devised [44]. It is calculated based on the following parameters: bilirubin, creatinine, INR, and the presence or absence of two dialysis sessions per week. The score ranges from 6 (stable) to 40 (which requires urgent transplantation). For HCC, in addition to a score generated by the calculation of the variables, an additional 0 and 22 points are awarded for a solitary tumor < than 2 cm, and solitary tumor 2-5 cm or 3 nodules each being <3 cm in size, respectively, and points are awarded for any additional 3 months on the waiting list for a transplant. In order to reduce the risk of progression of tumor while on the waiting list and subsequent dropout, evidence suggests that using living donors for transplantation is an alternative strategy. The 5-year survival is similar to that observed in cadaveric donors. Living donor transplantation is a cost-effective approach in reducing dropout while waiting on the transplant list, if this period exceeds 7 months [45]. The use of chemotherapy preoperatively to reduce tumor load while waiting on the list is controversial, because in addition to unclear benefits, there is a risk of seeding of the tumor during the procedure. Therefore, at this stage, recommendations are for use in highly selected patients on the waiting list for >6 months who have no peripheral tumor that can be the nidus for seeding. The options are transarterial chemoembolization, or for small tumors ethanol injection or radiofrequency ablation.

SURGICAL RESECTION

Surgical resection is the preferred treatment modality for solitary HCC (typically <5 cm) in patients without cirrhosis, because the risk of postoperative liver failure is lower than that in patients with cirrhosis. The 5-year survival after resection is at least 50% [46]. The best predictors of a good outcome after surgery are the bilirubin level and a hepatic vein pressure gradient (HVPG) <10 mmHg [47]. Clinical indicators of portal hypertension, i.e., varices, need for diuretic therapy, etc. are sufficient to preclude resection without the need for hepatic vein catheterization. Disappointingly, resection is associated with a high rate of recurrence, >70% at 5 years [48], both due to tumor spread and new tumor growth. At this stage, there are no effective or well-studied therapies for preventing or treating recurrence. In de novo tumors, re-resection might be an option or more ideally salvage transplantation, in carefully selected patients.

PERCUTANEOUS ABLATION

Percutaneous ablation may be offered as the therapeutic option of choice [49] but in most centers is used in patients who have early disease, but are not suitable for resection or liver transplantation. The best-studied method is ethanol injection of the tumor, usually done under ultrasound guidance. The smaller the tumor, the more effective ethanol injection is in inducing necrosis of tumor. In tumors <2 cm, 2-3 cm, and 3-5 cm, the rate of necrosis is up to 100%, 70%, and 50%, respectively [50–52]. In order to achieve complete necrosis, repeated injections are usually required; however, recurrences are typical. Post-treatment follow up is assessed by CT or MRI scanning after 1 month initially, and every 3–4 months thereafter. The radiological features are the absence of contrast uptake in the necrotic lesion; therefore, any contrast in the lesion is evidence of treatment failure. Radiofrequency ablation (RFA) is achieved by inserting single or multiple electrodes into the tumor mass and applying heat to cause necrosis of tissue. This technique works well for small lesions (<2 cm), but has superior performance for lesions >2 cm, when compared to ethanol injection [53]. Compared to ethanol injection, RFA is associated with significant morbidity (10%) and mortality (0.3%) [32]. The complications associated with RFA are pleural effusions, peritoneal bleeding, and risk of seeding (in poorly differentiated tumors). RFA has better survival data compared to ethanol injection [53] and may have comparable rates to resection [52]. The data support that RFA should not be used as first-line therapy in patients that are suitable for resection. Both RFA and ethanol are particularly useful as a bridge to transplantation, in patients with early HCC.

TRANSARTERIAL EMBOLIZATION AND CHEMOEMBOLIZATION

The rationale for arterial embolization is the dependence of HCC on blood supply for growth and spread. Thus, the hepatic artery may be embolized using a catheter, in a procedure called transcatheter arterial embolization (TAE). When this procedure is combined with injection of chemotherapy, it is referred to as transcatheter arterial chemoembolization (TACE). TAE and TACE are indicated in patients who have unresectable disease that is not amenable to ablation therapy, but who do not have extrahepatic metastases. The main contraindication is venous thrombosis or restricted venous flow. Occlusion of the blood flow induces necrosis in the tumor in >50% of patients [40]. TACE, however, should be used with caution in patients with advanced liver disease (Child Pugh B, C) because of the risk of liver failure and death [32]. Survival rates using this modality are between 20 and 60% at 2 years follow-up, which is an improvement compared to those not treated [54].

SORAFENIB

This is a multikinase inhibitor of several pathways, involved in tumor differentiation and growth, e.g., VEGFR2, PDGFR, Raf1, and others. Its use is indicated in patients who present with advanced disease or have failed TACE but with preserved liver function (ideally Child Pugh A). Sorafenib has been shown to increase life expectancy by over 2 months and time to progression of tumor compared to placebo [55].

OTHER TREATMENTS

Treatment with agents such as tamoxifen, antiandrogens, radioembolization with labeled beads, systemic and/or intra-arterial chemotherapy are not recommended, because the risk benefit ratio is unclear.

CONCLUSIONS

Viral hepatitis is a neglected cause of significant mortality and morbidity in Africa. With the availability of treatments, which cure HCV infection, and with improved access to HBV vaccination, there is the opportunity to dramatically reduce the incidence of HCC in Africa. Commitment by politicians, policy makers, public health experts, and academics is needed to address what could be one of the most effective health care interventions to prevent cancer in the next few decades.

TAKE HOME MESSAGES

- HCC is the third leading cause of cancer deaths.
- In SSA HCC tends to occur at a younger age, compared to western societies.
- In addition to the usual risk factors, HIV is associated with a 3-fold increase in HCC.
- The most common symptom and clinical sign is RUQ pain and hepatomegaly, respectively.
- Diagnosis is made by typical radiological features, however in SSA AFP is still a useful albeit insensitive and nonspecific diagnostic test.
- Biopsy is only recommended for tumors with atypical radiological features. Treatment options for early stage disease are transplantation, resection, and percutaneous ablation.
- Prognosis for BCLC stage 0/A is good; therefore, early diagnosis and referral to a tertiary center with an MDT approach is essential.
- Treatment for advanced disease is TACE, sorafenib, or supportive care.
- Living donor transplantation is a cost-effective approach with a 5-year survival similar to that observed in cadaveric donors.

REFERENCES

[1] Global Burden of Disease Liver Cancer Collaboration. The burden of primary liver cancer and underlying etiologies from 1990 to 2015 at the global, regional, and national level: results from the global burden of disease study 2015. JAMA Oncol 2015;3:1683.

[2] Ferlay J, et al. Estimates of worldwide burden of cancer in 2008: GLOBOCAN 2008. Int J Cancer 2010;127(12):2893–917.

[3] Yang JD, Roberts LR. Hepatocellular carcinoma: a global view. Nat Rev Gastroenterol Hepatol 2010;7(8):448–58.

[4] El-Serag HB, Kanwal F. Epidemiology of hepatocellular carcinoma in the United States: where are we? Where do we go? Hepatology 2014;60(5):1767–75.

[5] Prates MD, Torres FO. A cancer survey in Lourenco Marques, Portuguese East Africa. J Natl Cancer Inst 1965;35(5):729–57.

[6] Kew MC, Geddes EW. Hepatocellular carcinoma in rural southern African blacks. Medicine 1982;61(2):98–108.

[7] Jemal A, et al. Global cancer statistics. CA Cancer J Clin 2011;61(2):69–90.

[8] Dhanasekaran R, Bandoh S, Roberts LR. Molecular pathogenesis of hepatocellular carcinoma and impact of therapeutic advances. F1000Res 2016;**5**:859.

[9] Chen CJ, Yang HI. Natural history of chronic hepatitis B REVEALed. J Gastroenterol Hepatol 2011;26(4):628–38.

[10] Saves M, et al. Risk factors for coronary heart disease in patients treated for human immunodeficiency virus infection compared with the general population. Clin Infect Dis 2003;37(2):292–8.
[11] Chasela CS, et al. Hepatitis B virus infection among HIV-infected pregnant women in Malawi and transmission to infants. J Hepatol 2014;60(3):508–14.
[12] Kew MC, et al. Increased hepatocarcinogenic potential of hepatitis B virus genotype A in Bantu-speaking sub-Saharan Africans. J Med Virol 2005;75(4):513–21.
[13] Menendez D, Shatz M, Resnick MA. Interactions between the tumor suppressor p53 and immune responses. Curr Opin Oncol 2013;25(1):85–92.
[14] Gouas DA, et al. Effects of the TP53 p.R249S mutant on proliferation and clonogenic properties in human hepatocellular carcinoma cell lines: interaction with hepatitis B virus X protein. Carcinogenesis 2010;31(8):1475–82.
[15] Zhang YJ, et al. Aflatoxin B1 and polycyclic aromatic hydrocarbon adducts, p53 mutations and p16 methylation in liver tissue and plasma of hepatocellular carcinoma patients. Int J Cancer 2006;119(5):985–91.
[16] Davila JA, et al. Hepatitis C infection and the increasing incidence of hepatocellular carcinoma: a population-based study. Gastroenterology 2004;127(5):1372–80.
[17] Mohamed AE, Kew MC, Groeneveld HT. Alcohol consumption as a risk factor for hepatocellular carcinoma in urban southern African blacks. Int J Cancer 1992;51(4):537–41.
[18] Umoh NJ, et al. Aetiological differences in demographical, clinical and pathological characteristics of hepatocellular carcinoma in the Gambia. Liver Int 2011;31(2):215–21.
[19] Donato F, et al. Alcohol and hepatocellular carcinoma: the effect of lifetime intake and hepatitis virus infections in men and women. Am J Epidemiol 2002;155(4):323–31.
[20] Beasley RP, Hwang LY. Hepatocellular carcinoma and hepatitis B virus. Semin Liver Dis 1984;4(2):113–21.
[21] Fattovich G, et al. Occurrence of hepatocellular carcinoma and decompensation in western European patients with cirrhosis type B. The EUROHEP study group on hepatitis B virus and cirrhosis. Hepatology 1995;21(1):77–82.
[22] Mandishona E, et al. Dietary iron overload as a risk factor for hepatocellular carcinoma in Black Africans. Hepatology 1998;27(6):1563–6.
[23] Hassan MM, et al. The association of family history of liver cancer with hepatocellular carcinoma: a case-control study in the United States. J Hepatol 2009;50(2):334–41.
[24] Shen FM, et al. Complex segregation analysis of primary hepatocellular carcinoma in Chinese families: interaction of inherited susceptibility and hepatitis B viral infection. Am J Hum Genet 1991;49(1):88–93.
[25] Thio CL, et al. HIV-1, hepatitis B virus, and risk of liver-related mortality in the Multicenter Cohort Study (MACS). Lancet 2002;360(9349):1921–6.
[26] Tanon A, et al. The spectrum of cancers in West Africa: associations with human immunodeficiency virus. PLoS One 2012;7(10):e48108.
[27] Gomez-Gonzalo M, et al. The hepatitis B virus X protein induces HIV-1 replication and transcription in synergy with T-cell activation signals: functional roles of NF-kappaB/NF-AT and SP1-binding sites in the HIV-1 long terminal repeat promoter. J Biol Chem 2001;276(38):35435–43.
[28] Kew MC, et al. Hepatitis-B antigen and cirrhosis in Bantu patients with primary liver cancer. Cancer 1974;34(3):538–41.
[29] Kew MC. Hepatocellular carcinoma in Sub-Saharan Africa. Ann Hepatol 2012;343.
[30] Gregory B, Ho VC. Cutaneous manifestations of gastrointestinal disorders, Part II. J Am Acad Dermatol 1992;26(3 Pt 2):371–83.
[31] Luo JC, et al. Clinical characteristics and prognosis of hepatocellular carcinoma patients with paraneoplastic syndromes. Hepatogastroenterology 2002;49(47):1315–9.
[32] Bruix J, Sherman M. Management of hepatocellular carcinoma: an update. Hepatology 2011;53(3):1020–2.
[33] European Association for the Study of the Liver; European Organisation for Research and Treatment of Cancer. EASL-EORTC clinical practice guidelines: management of hepatocellular carcinoma. J Hepatol 2012;56(4):908–43.

[34] Burrel M, et al. MRI angiography is superior to helical CT for detection of HCC prior to liver transplantation: an explant correlation. Hepatology 2003;38(4):1034–42.
[35] Roskams T, Kojiro M. Pathology of early hepatocellular carcinoma: conventional and molecular diagnosis. Semin Liver Dis 2010;30(1):17–25.
[36] Di Tommaso L, et al. Diagnostic value of HSP70, glypican 3, and glutamine synthetase in hepatocellular nodules in cirrhosis. Hepatology 2007;45(3):725–34.
[37] Capurro M, et al. Glypican-3: a novel serum and histochemical marker for hepatocellular carcinoma. Gastroenterology 2003;125(1):89–97.
[38] International Consensus Group for Hepatocellular Neoplasia. Pathologic diagnosis of early hepatocellular carcinoma: a report of the international consensus group for hepatocellular neoplasia. Hepatology 2009;49(2):658–64.
[38a] Llovet JM, et al. The Barcelona approach: diagnosis, staging, and treatment of hepatocellular carcinoma. Liver Transpl 2004;10(2 Suppl. 1):S115–20.
[39] Roayaie S, et al. Resection of hepatocellular cancer </=2 cm: results from two western centers. Hepatology 2013;57(4):1426–35.
[40] Llovet JM, Bruix J. Systematic review of randomized trials for unresectable hepatocellular carcinoma: chemoembolization improves survival. Hepatology 2003;37(2):429–42.
[41] Llovet JM, Bruix J. Molecular targeted therapies in hepatocellular carcinoma. Hepatology 2008;48(4):1312–27.
[42] Llovet JM, Bru C, Bruix J. Prognosis of hepatocellular carcinoma: the BCLC staging classification. Semin Liver Dis 1999;19(3):329–38.
[43] Mazzaferro V, et al. Liver transplantation for the treatment of small hepatocellular carcinomas in patients with cirrhosis. N Engl J Med 1996;334(11):693–9.
[44] Kamath PS, et al. A model to predict survival in patients with end-stage liver disease. Hepatology 2001;33(2):464–70.
[45] Sarasin FP, et al. Living donor liver transplantation for early hepatocellular carcinoma: a life-expectancy and cost-effectiveness perspective. Hepatology 2001;33(5):1073–9.
[46] Grazi GL, et al. Improved results of liver resection for hepatocellular carcinoma on cirrhosis give the procedure added value. Ann Surg 2001;234(1):71–8.
[47] Bruix J, et al. Surgical resection of hepatocellular carcinoma in cirrhotic patients: prognostic value of preoperative portal pressure. Gastroenterology 1996;111(4):1018–22.
[48] Shirabe K, et al. Factors linked to early recurrence of small hepatocellular carcinoma after hepatectomy: univariate and multivariate analyses. Hepatology 1991;14(5):802–5.
[49] Sato S, et al. Power Doppler signals after percutaneous ethanol injection therapy for hepatocellular carcinoma predict local recurrence of tumors: a prospective study using 199 consecutive patients. J Hepatol 2001;35(2):225–34.
[50] Okada S. Local ablation therapy for hepatocellular carcinoma. Semin Liver Dis 1999;19(3):323–8.
[51] Ishii H, et al. Local recurrence of hepatocellular carcinoma after percutaneous ethanol injection. Cancer 1996;77(9):1792–6.
[52] Livraghi T, et al. Percutaneous ethanol injection in the treatment of hepatocellular carcinoma in cirrhosis. A study on 207 patients. Cancer 1992;69(4):925–9.
[53] Cho YK, et al. Systematic review of randomized trials for hepatocellular carcinoma treated with percutaneous ablation therapies. Hepatology 2009;49(2):453–9.
[54] Bruix J, Sala M, Llovet JM. Chemoembolization for hepatocellular carcinoma. Gastroenterology 2004;127(5 Suppl 1):S179–88.
[55] Llovet JM, et al. Sorafenib in advanced hepatocellular carcinoma. N Engl J Med 2008;359(4):378–90.

CHAPTER

BILIARY DISEASE IN SUB-SAHARAN AFRICA

17

M.D. Smith*, **J.W.S. Devar**†

*Academic Head of Surgery, University of Witwatersrand, Johannesburg, South Africa**

Chris Hani Academic Hospital and University of the Witwatersrand, Johannesburg, South Africa†

INTRODUCTION

Biliary tract disorders are common in sub-Saharan Africa (SSA). The limited resources and poor socio-economic environment makes this group of diseases extremely challenging to investigate and treat. The wide spectrum of disease includes parasitic infections, HIV disease, and common diseases that affect HICs, i.e., gall stone disease. This chapter focuses mainly on the common diseases and highlights our experience.

GALLSTONES (CHOLECYSTOLITHIASIS)

Historically, gallstones were rare in sub-Saharan Africa (SSA). Although there are no accurate epidemiological data on the performance of cholecystectomies, it was not a common procedure in SSA. In 1927, Beyers [1] found that cholecystic disease was found in only four cases among 18,000 black patients at the Johannesburg General Hospital and again in 1934 Brebner [2] working at the same Johannesburg Hospital found that only 0.67 per 1000 patients were diagnosed with gallbladder disease. However, over the last two decades, the number of cholecystectomies performed has increased dramatically and it now probably represents one of the more common operations carried out at Chris Hani Baragwanath Academic Hospital (CHBAH),where in excess of 300 cholecystectomies per year are done mostly laparoscopically. This is compared to 118 over 3 years as reported by Parek in 1987 [3].

The precise reason for this increase is not well defined. It is thought to be due to urbanization and an increased intake of a western-type diet—especially the availability of cheap poor-quality fast foods. Walker and Segal et al. reported on a series of 100 patients in 1985 who were screened for gallstones and 10% of women in the study were found to have stones. They suggested that an increase in incidence could be expected in the future [4]. In our experience, this has occurred and we have noted an increase in incidence of Cholecystolithiasis in both females and males. Pure pigment stones are uncommon and there has not been a dramatic increase due to sickle cell anemia nor the blood dyscrasias that lead to their formation.

Digestive Diseases in Sub-Saharan Africa. https://doi.org/10.1016/B978-0-12-815677-3.00017-7

GALLSTONE PANCREATITIS

Linked to the increase in gallstone prevalence is an increase in the complications associated with gallstones. There has been a dramatic increase in the prevalence of gallstone pancreatitis. This type of pancreatitis was rare but it is rapidly becoming common in both females and in males. In addition, the age of patients presenting with pancreatitis has increased. The number of patients presenting with severe acute pancreatitis (SAP) with end-organ dysfunction requiring surgical interventions has increased.

Pancreatitis should initially be managed conservatively with aggressive organ support. The recently modified Atlanta classification identifies three levels of severity—mild, moderately severe, and severe [5]. Moderately severe disease implies organ dysfunction that reverses with organ support in 48 h, while severe pancreatitis implies progressive organ dysfunction. The moderately severe and severe groups should be managed in the ICU but *because of limited resources these patients have to be managed in general wards*. The management requires interventional approaches to infected walled off necrosis (WON), which is complex and relies on the availability of significant skills and resources. The current recommended approach is firstly to delay surgical intervention as long as possible and, if it is required, to use the "step up" approach, which requires a *multidisciplinary team including a radiologist, gastroenterologist surgeon, and critical care intensivist* [6]. The approach to the infected WON is determined by the location of the infected collection—either by a minimally invasive retroperitoneal necrosectomy with access via the left flank or endoscopically utilizing a EUS placed stent followed by endoscopic debridement on demand (Fig. 17.1).

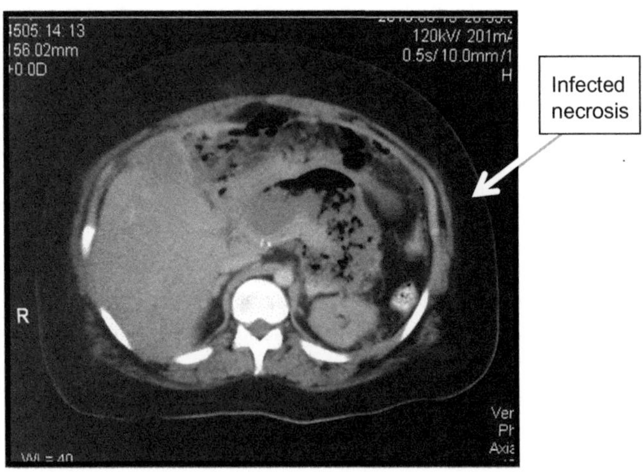

FIG. 17.1

Infected walled-off necrosis.

CHOLEDOCHOLITHIASIS

Obstructive jaundice due to gallstones in the common bile duct has increased. In hospitals where ERCP (endoscopic retrograde cholangio-pancreatography) is available, this is the modality of choice for treatment. This minimal access approach is highly successful with low complications rates. *Unfortunately, there are few hospitals performing this procedure, and therefore patients are subject to open laparotomy for the clearance of the duct stones. Advanced laparoscopy is still in its infancy and there are very few centers that would consider laparoscopic common bile duct explorations.*

Cholangitis is a common presenting problem with obstructive jaundice (OJ). The diagnosis is often missed and remains a clinical diagnosis with biochemical confirmation. *Due to delays in presentation, patients often present with a severe form of cholangitis and have organ dysfunction, most commonly renal dysfunction. Limited access to renal replacement therapy makes management of this condition very difficult.* The use of nephrotoxic drugs must be avoided and this includes the use of nonsteroidal drugs for pain management. As the age of the patients increases, so does the comorbidities. Hypertensive heart disease and all its sequelae are not uncommon and therefore organ dysfunction can be a terminal event. Patients can be stratified into a mild and severe type, based on clinical (Reynaud's Pentad) and biochemical parameters. The severe form of the disease implies organ dysfunction and as such this subset of patients should be managed in an *appropriate setting, preferably a high care unit.* We recommend early transfer of all patients to an appropriate center that can deal with the obstruction. If there has not been an adequate clinical response within 24–36h, decompression of the biliary system is indicated. We prefer ERCP as our initial intervention; however, patients with severe cholangitis who are unfit for an ERCP should have a percutaneous biliary drainage (PBD). The morbidity rates of PBD remain high and the interventional radiology skills are often limited.

Most patients without severe cholangitis will respond to antibiotics and can then be treated electively by open surgery and common bile duct exploration.

Mirizzi's syndrome is a condition where a stone impacted in Hartman's pouch compresses the common hepatic duct resulting in OJ. There are a number of grades of the condition from extrinsic compression to fistularization between the common hepatic duct (CHD) and the gallbladder (Fig. 17.2).

It is recommended that, in general, a Mirizzi's syndrome is an indication for open cholecystectomy and therefore it is important to diagnose this condition preoperatively. The radiological appearance on the cholangiogram suggests extrinsic compression and further cross-sectional imaging may be required to exclude other conditions such as gallbladder cancer—a pathology, which is becoming more common in the region.

BILE DUCT INJURIES

The increased prevalence of symptomatic gallstones implies a greater number of cholecystectomies and hence a greater number of postoperative bile leaks and iatrogenic bile duct injuries. These are usually complex injuries and *require centers of excellence* to manage them [7]. A significant number of bile duct leaks are due to cystic duct stump leaks, associated at times with retained common bile duct stones. This group is easily managed by ERCP, stone removal, and an endobiliary stent to manage the leak. *Availability of ERCP is an important issue that needs to be addressed in SSA. As the prevalence of stones increases, so does the necessity for these minimal access procedures increase.*

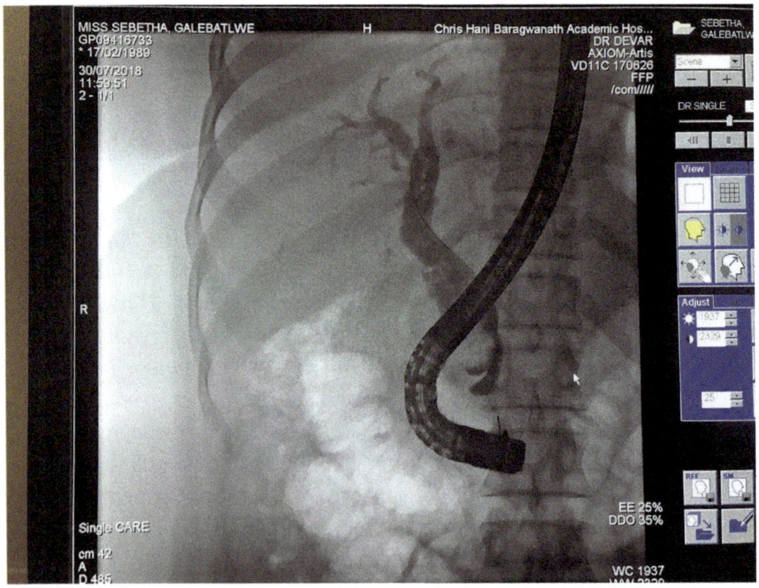

FIG. 17.2

Mirizzi's syndrome with low insertion of cystic duct.

A good knowledge of how to manage CBD complications is essential to prevent significant morbidity and even mortality in these patients. If recognized at the time of surgery, a drain should be placed in the CBD *and patients referred to a center where the reconstruction can be performed.* Where the injury is not recognized, a high level of suspicion should be exercised in the immediate postoperative period where patients should make a rapid recovery. Early sonar to identify free fluid in the peritoneal cavity is essential. If this is identified, a rapid peritoneal wash out is required with the placement of the drain *and early referral to a high volume center. There is evidence to support referral to a center with the necessary skills* [7]. The first attempt at reconstruction is always the most important and likely to succeed. There are data to support delayed repair and so while this may result in some patient discomfort the results are the best [8].

BILE DUCT CYSTS

Bile duct cysts are mainly a problem of childhood; however, diagnosis is delayed in nearly 20% of patients. Adults have an increased rate of complications and they may present with complications of previous procedures to treat the cysts. There is consensus that excision is the appropriate surgical management of the extrahepatic bile duct cysts.

The most widely accepted hypothesis is that cystic dilatation is related to an anomalous arrangement of the pancreaticobiliary junction, with a long common channel that allows reflux of pancreatic juice into the biliary tree causing inflammation, ectasia, and ultimately dilatation [9]. This long common channel is frequently noted in our patients presenting with bile duct cysts.

Other theories of the etiology include Sphincter of Oddi dysfunction, hereditary factors, and aganglionosis in the distal neck.

The most common type of bile duct cyst that we see is the Type I as classified by Todani (Fig. 17.3).

In our population group, the presenting symptoms are mainly due to recurrent epigastric pain, intermittent jaundice, and cystolithiasis. Up to 20% may present with biliary pancreatitis. We have also had patients presenting with portal hypertension and cavernous transformation of the portal vein.

Imaging plays a vital role in the diagnosis and management of these cysts and *cross-sectional imaging using CT scans and MRI with MRCP is readily used.*

Malignancies arising from the hepatobiliary tract and associated with bile duct cysts has ranged from 2.5% to 28%. Malignancies recorded are cholangiocarcinoma, squamous cell carcinoma, anaplastic carcinoma, bile duct sarcoma, pancreatic carcinoma, and gallbladder carcinoma.

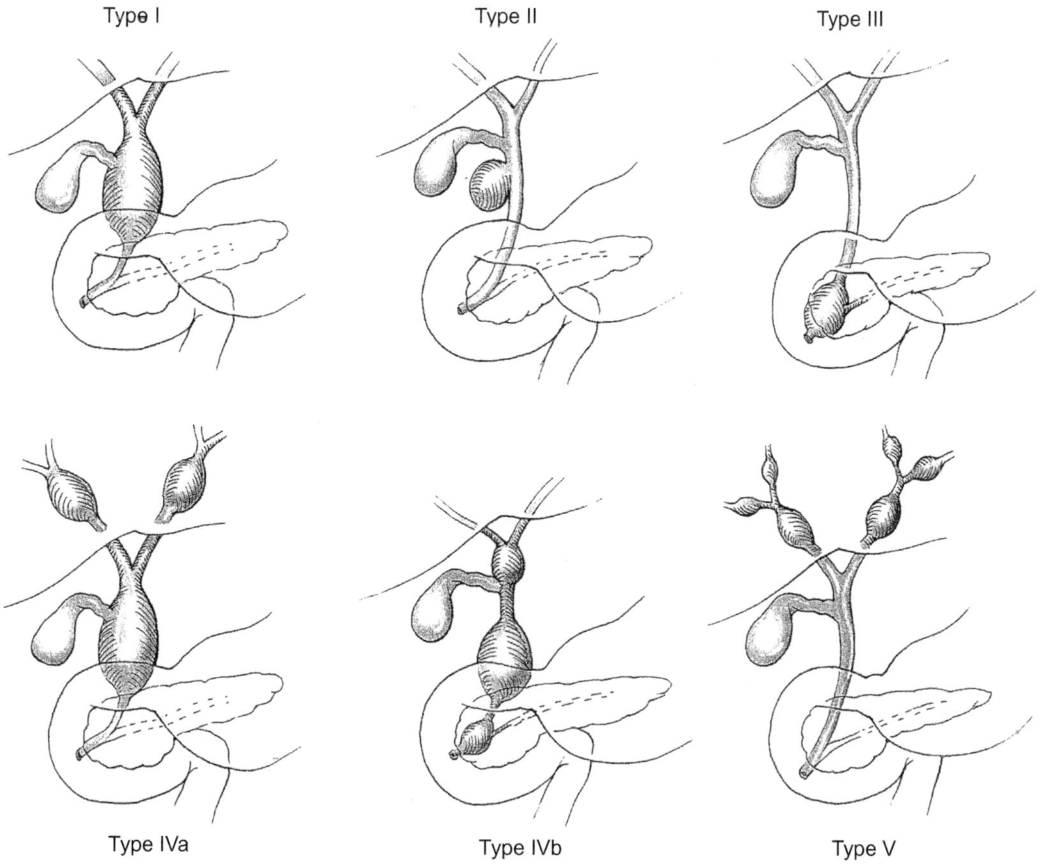

FIG. 17.3

Todani classification of bile duct cysts.

Cholangiocarcinoma is the most frequent malignancy associated with bile duct cysts. Its incidence is 20–30 times greater than the general population. The risk of malignancy remains high even after resection of the bile duct cyst.

Initial treatment of bile duct cysts requires control of biliary infections and in patients where sepsis fails to respond to intravenous antibiotics, preoperative percutaneous or endoscopic decompression of the infected bile duct cyst is required.

The definitive management requires excision of the cyst and mucosa to mucosa biliary-enteric anastomosis. If complete excision is not possible, then partial excision with obliteration of the remaining "at risk" mucosa and Roux-en-Y cystojejunostomy is preferred.

GALLBLADDER CANCER

Gallbladder adenocarcinoma is reported as being the most common biliary tract cancer. The etiology is uncertain, but in some countries it is linked to the presence of gallstones. In the Indian subcontinent, there is an exposure to carcinogens such as the water in the Ganges river. The *etiology in SSA is unclear*. Gallstones are present in many of our patients. Findings of incidental carcinoma in cholecystectomy specimens are not common. The majority of patients present with OJ and have advanced disease. Liver metastasis is common as is the involvement of celiac, para-aortic, and retropancreatic lymph nodes. This renders the majority of lesions of irresectable. However, OJ alone is not a contraindication to resection. In patients with a T2 (contained within the gallbladder wall) lesion, an extended cholecystectomy and portal lymphadenectomy is performed. T3 (extending beyond the gallbladder wall) lesions and above are usually irresectable. However, we have performed extended resections for the more advanced lesions with good results. *This relatively uncommon cancer is becoming more common in our practice*. We are unsure of the reasons for this and are currently exploring some theories but suspect that they are clustered in certain geographical regions and so the exposure to carcinogens as seen in the Indian patients may be present in our region. In 1995, Warren et al. reported on 19 cases over a 68-month period [10]. *At CHBAH, we see this number in 3 months. They also reported a regional variation suggesting this is an important aspect to investigate further.*

The diagnosis usually requires advanced cross-sectional imaging for diagnosis and staging. Tissue is obtained from the most convenient site and the relief of OJ is based on the level of the obstruction in the irresectable patient. In obstructions below the confluence of the left and right hepatic ducts, we favor the use of ERCP and stenting. However, when there is involvement of the confluence, we would use the PBD approach.

Palliation includes the relief of OJ and we use a combination of Gemcitabine and cisplatinum to promote stabilization of the disease. On rare occasions, down staging of the disease occurs, but the type of resection performed is based on the original cross-sectional imaging.

CHOLANGIOCARCINOMA

This is a rare tumor occurring within the sixth or seventh decade of our population, but when it does present the majority of our patients have advanced disease that requires palliation of the obstructive jaundice and palliative chemotherapy. In the last 3 years at CHBAH, we have only managed to take one patient with this disease to resection. Most tumors are located in the peri-hilar area and our approach to biliary drainage depends on their Bismuth-Corlette classification. Those with type I and Type II lesions are approached endoscopically, while types III and IV are approached with percutaneous biliary drainage. The palliative chemotherapy is once again gemcitabine and cisplatinum regimen, as supported by the ABC-02 trial [11].

i) Parasitic Infections of the Biliary tree

Intestinal parasites are the most common micro-organisms to affect humans. It is estimated that approximately 25% of the LICs population is infested with Ascaris lumbricoides [12]. Biliary parasites can cause biliary obstruction, biliary lithiasis, hemobilia, sclerosing cholangitis, and cholangiocarcinoma. The stones related to ascaris are thought to form in the gallbladder and not the ducts [13]. Our population group is not affected by the liver flukes (*Fasciola hepatica*, Chlonorchis sinensis, and *Opisthorchis viverrini*). Ascaris may also migrate into the pancreatic duct and cause acute pancreatitis particularly in the pediatric age group.

ii) HIV and AIDS

The incidence of HIV and AIDS has not significantly decreased in SSA. The introduction of HAART has changed the spectrum of the disease. The incidence in a surgical department remains at overall 30% with a higher incidence among young females [14]. There are very few data regarding biliary diseases in these immunocompromised patients. It has been reported that 50% of patients will have abnormal liver function tests during the course of their disease. In addition, many of the antiretroviral drugs have an effect on the liver aggravating the abnormal liver functions.

In HIV and AIDS, the patients present with pathology directly related to the immunocompromised patient or they present with pathology unrelated but affected by the immune status. Furthermore, we see patients who present with the complications of HIV and AIDS therapy. Right upper quadrant pain as a consequence of gallbladder pathology has been well described. The incidence of acalculous cholecystitis was reported from the west coast of the USA [15]. The diagnosis and management was described and cholecystectomy was recommended. The morbidity and mortality was slightly higher in these patients. Cholecystectomy was recommended for patients with acalculous cholecystitis. This has not been our experience. While many patients with symptomatic gallstones do undergo cholecystectomy, the acalculous variant is not well described in our patient population group.

AIDS-related cholangiopathy was described and classified by Cello in the late 1980s [16]. This form of secondary sclerosing cholangitis is difficult to distinguish from primary sclerosing cholangitis (PSC). There is some suggestion that it is predominantly an extra hepatic form of PSC. The etiology of AIDS cholangiopathy is unknown, but is believed to be related to opportunistic infections of the biliary tract where pathogens have often been detected. These infections likely cause a secondary sclerosing cholangitis due to the associated continual inflammation. It is theorized that enteric infection leads to portal bacteremia, which, in turn, leads to subsequent bile duct injury and destruction [17]. HIV itself may play a role in some manifestations of AIDS cholangiopathy. However, although it has been isolated in the bowel mucosa, it has never been detected in the biliary epithelium. A genetic predisposition has been postulated as pathogens may trigger an autoimmune-mediated reaction in patients with HLA DRw52a [17]. This may explain why features of AIDS cholangiopathy are independent of the associated opportunistic infection [18].

The following table describes some of the opportunistic organisms implicated.

Bacteria	Viruses	Protozoa	Fungi
Mycobacterium avium-intracelluare	Cytomegalovirus	*Pneumocystis carinii*	*Histopasma capsulatum*
Mycobacterium kansasii	Herpes simplex	*Microsporidia*	*Cryptococcus neoformans*
Mycobacterium tuberculosis	Adenovirus	*Cryptosporidia*	*Candida albicans*
Salmonella enteritidis	HIV	*Leishmania donovani*	*Cociious immitis*
Salmonella typhimurium		*Toxoplasma gondii*	
Campylobacter fetus		*Isospora*	
Enterobacter cloacae			

Clinically, the patients present with abnormal liver function tests and may be biochemically and clinically jaundiced. The entity was originally described in patients with low CD4 counts <200. However, in our experience, we have seen patients with the clinical and cholangiographic picture but with CD4 counts >200. Is ARC an AIDS-defining disease? There was previously a suggestion that it should be, but our experience with patients suggests that this may not be the case. The diagnosis is made based on clinical suspicion but confirmed on the cholangiographic appearance. In up to 40%, there are concomitant abnormalities in the pancreatic duct, which are usually asymptomatic.

There are no pathognomonic histopathological findings and as such there can be little reliance on the pathology report. Cello described 4 types-papillary stenosis and cholangitis (most common presentation, 50%), papillary stenosis alone (30%), intrahepatic sclerosing cholangitis alone (10%), long extrahepatic bile duct stricture (10%) [15].

Treatment is difficult and is aimed at symptomatic relief. Sphincterotomy of the papillary stenosis subtype is very successful. Balloon dilatation is helpful in local stricturing. Stenting is only of value for patients with cholangitis and then the stents should be removed as soon as possible. Ursotan is used in symptomatic patients with good response, but the long-term prognosis using deoxycholic acid is unknown.

The response to ARVs is not well defined. However, what is not clear is whether there can be any resolution of the sclerosing pathology in a response to these drugs. With huge ARV coverage in SA, we still see ARC and so the response to therapy needs to be evaluated.

OBSTRUCTIVE JAUNDICE

The majority of patients presenting at the first point of contact with the health services with jaundice have not got obstructive jaundice (OJ). However, patients with OJ have a higher morbidity and mortality rate and as such OJ must be excluded in all patients presenting with jaundice before they are managed according to protocols for prehepatic and hepatic causes of the jaundice. The diagnosis is usually suggested clinically, but the three cardinal clinical symptoms of jaundice, dark urine, and pale stools are not always present. After a thorough clinical history and examination, urinary dipsticks looking for bilirubin in the urine is helpful in suggesting OJ. The value of the liver function tests is well described, but the availability is not always timeous nor constant. Sonar to detect a dilated biliary tree is very helpful but again not always available. As such, it is important that the approach is to exclude OJ before a supportive care protocol is implemented.

The second important consideration is to exclude complications of OJ. The three main complications from OJ include cholangitis (discussed earlier), renal dysfunction, and coagulopathy. Cholangitis is a clinical diagnosis. Renal dysfunction is difficult to diagnose clinically, but any cause of increased fluid loss such as fever or vomiting, or decreased intake such as confusion should alert the clinician to the possible risk of renal dysfunction. The most important aspects of management include avoiding nephrotoxic drugs, rapid and adequate rehydration, and the management of cholangitis. Unlike the renal failure associated with the hepatorenal syndrome, this is a form of acute tubular necrosis and can be reversible in the majority of patients. As such, a nihilistic approach of withholding renal replacement therapy is inappropriate especially when the cause of the OJ is benign disease. Bleeding caused by the coagulopathy is seldom the presenting problem. As such, the real relevance of the coagulopathy is related to the high likelihood of an invasive procedure required during the management of these patients. In the majority of patients, the liver synthetic function remains intact and the problem is the absorption of the fat-soluble vitamin K from the GIT. Systemic administration of Vit K results in a rapid reversal of the coagulopathy. However, if an invasive procedure is required in the first 48–72 h, it is recommended that fresh frozen plasma is administered to the patient with an INR > 1.4.

After the confirmation of OJ and the assessment of complications, the search for the etiology becomes important. The evaluation of the underlying pathology usually requires advanced investigations including sonar, cross-sectional imaging including CT or MRI, MRCP and today EUS is very helpful but its availability is limited to major centers. ERCP is no longer considered part of the diagnostic armamentarium and its value has been clearly identified as therapeutic. This is due to the considerable morbidity (10%) associated with ERCP. Less invasive procedure will provide the same information as the cholangiogram obtained by ERCP.

The current recommended approach is to resuscitate and stabilize the patient, obtain the aforementioned imaging prior to any decompression of the biliary system unless the patient is cholangitic or has renal dysfunction. The further management will be based on the etiology and prompt referral to a center, which has the necessary resources and skills, to complete the definitive management.

CONCLUSIONS

Biliary tract disorders are common in sub-Saharan Africa. There are limited databases that can accurately reflect the spectrum of disease, but the extent of the pathology is well highlighted within this chapter. The late presentation of patients increases the morbidity and mortality and further expertise is currently being developed to address the shortfalls outside centers of excellence.

REFERENCES

[1] Beyers CF. Incidence of surgical diseases among the Bantu races of South Africa. S Afr Med J 1927;1:606–12.

[2] Brebner IW. Surgical conditions the liver, including the gallbladder. S Afr Med J 1934;8:867–71.

[3] Parekh D, Lawson HH, Kuyl JM. Gallstone disease among black South Africans. A review of the Baragwanath hospital experience. S Afr Med J 1987;72(1):23–6.

[4] Walker AR, Segal I, et al. Prevalence of gallstones in elderly black woman in Soweto, Johannesburg, as assessed by ultrasound. Am J Gastroenterol 1989;84:1383–5.

[5] Banks PA, et al. Classification of acute pancreatitis-2012: revision of the Atlanta classification and definitions by international consensus. Gut 2013;62:102–11.

[6] Van Santvoort HC, Bakker OJ, Bollen TL, et al. A conservative approach to necrotising pancreatitis improves outcomes. Gastroenterology 2011;141:1254–63.
[7] Mercado MA, Franssen B, Dominguez I, et al. Transition from a low: to a high-volume centre for bile duct repair: changes in technique and improved outcome. HPB (Oxford) 2011;13(11):767–73.
[8] Iannelli A, Paineau J, Hamy A, et al. Primary versus delayed repair for bile duct injuries sustained during cholecystectomy: results of a survey of the association Francaise de Chirurgie. HPB (Oxford) 2013;15(8):611–6.
[9] Babbit DP. Congenital choledochal cysts: new etiological concept based on anomalous relationships of common bile duct and pancreatic bulb. Ann Rad 1969;12:231–40.
[10] Warren BL, Marais AW. Carcinoma of the Gallbladder—a possible regional predisposition in the western Cape and northern Cape. S Afr J Surg 1995;33(4):161–4.
[11] Valle J, Wassan H, et al. Cisplat plus gemcitabine versus gemcitabine for biliary tract cancer. NEJM 2010;362:1273–81.
[12] World Health Organization. Control of Ascaris. A report of WHI expert committee with technical rep. Ser. No. 379, Geneva: WHO; 1967.
[13] Yellin AE, Donovan AJ. Biliary lithiasis and helminthiasis. Am J Surg 1981;142:128–36.
[14] Martinson NA, Omar T, Gray GE, Vermaak JS, Badicel M, Degiannis E, Steyn J, McIntyre J, Smith MD. High rates of HIV in surgical patients in Soweto, South Africa: impact on resource utilization and recommendations for HIV testing. Trans R Soc Trop Med Hyg 2007;10(2):176–82.
[15] Cello JP. AIDS-related biliary tract disease. Gastrointest Endosc Clin N Am 1998;8(4):963–73.
[16] Cello JP. Acquired immunodeficiency syndrome cholangiopathy: spectrum of disease. Am J Med 1989;86(5):539–46.
[17] Nash JA, Cohen SA. Gallbladder and biliary tract disease in AIDS. Gastroenterol Clin North Am 2001;26:79–81.
[18] Kumar KS, Nair S, et al. Isolated intrahepatic biliary dilatation in a patient with acquired immune deficiency syndrome(AIDS): Aids cholangiopathy versus incidental unilobar Caroli's disease. J Clin Gastroenterol 2001;32:79–81.

CHAPTER

AMOEBIASIS IN SUB-SAHARAN AFRICA

18

Mashiko Setshedi*, S.R. Rutherford Thomson†,‡

Department of Medicine, University of Cape Town, Cape Town, South Africa St. Anne's Trust Chair and Head of Division of Medical Gastroenterology, Department of Medicine, University of Cape Town, Cape Town, South Africa†*
University College, London, United Kingdom‡

INTRODUCTION

Amoebiasis refers to a parasitic infection caused by the protozoan *Entamoeba histolytica* (*E. histolytica*). Although there are other species in the genus Entamoeba, only *E. histolytica* is associated with disease in humans. While *E. histolytica* has worldwide prevalence, it occurs most commonly in tropical areas. This is largely due to poor sanitation and the unavailability of running water in some cases. *E. histolytica* is spread via the feco-oral route and has potent tissue-lysing capabilities. Infection with *E. histolytica* may be asymptomatic, or cause invasive disease with dysentery and extraintestinal disease. Notably *E. histolytica* is associated with significant mortality with an estimated 100,000 deaths *per annum* [1]. The diagnosis and management of this condition has significantly improved outcomes despite the HIV epidemic, but when these are delayed there remains significant mortality. In this chapter, we present amoebiasis in the context of Sub-Saharan Africa and discuss how clinical and scientific research in the region have contributed to solutions to a variety of clinical challenges in the diagnosis, management, and complications of this disease.

EPIDEMIOLOGY

The stated prevalence of *E. histolytica* worldwide ranges from 10% to 50% [2, 3]. *E. histolytica* occurs mainly in Africa, India, Far East, Mexico, South and Central America. These figures, however, were calculated before the availability of molecular techniques enabling the separation of the nonpathogenic *E. dispar* from *E. histolytica*. In fact the first comprehensive study to characterize the zymodeme pattern of the amoeba, leading to a differentiation between *E. histolytica* and *E. dispar*, was conducted in 1982 in South Africa by Sergeaunt et al. [4]. Consequently, it is likely that the true prevalence of *E. histolytica* is lower than previously stated. In Sub-Saharan Africa, prevalence studies are lacking, making estimations difficult; nonetheless, the prevalence of *E. dispar* is higher than that of *E. histolytica* as in other parts of the world. In a South African study, 90% of asymptomatic individuals were infected with *E. dispar* while 10% asymptomatically harbored *E. histolytica* [5]. In another South African cohort with amoebic liver abscess, the overall *E. histolytica* and *E. dispar* prevalence was 2.4% and 8%,

Digestive Diseases in Sub-Saharan Africa. https://doi.org/10.1016/B978-0-12-815677-3.00018-9

respectively, over a 5-year period [6]. The prevalence varies according to the type of test and sample used to diagnose disease. Both stool culture and stool PCR yielded a prevalence of between 1.2 and 2% in South Africa, Egypt, and Ivory Coast in asymptomatic individuals [7]. In the same populations, however, when testing for serum antibodies, the seroprevalence rates were much higher, i.e., 6.9%–20.5%. In the developing world, *E. histolytica* occurs more commonly in migrants and returning travelers who have spent more than one month in an endemic area. In addition, in the USA and Europe, HIV infection is not a risk factor for invasive disease [8] and homosexual men are predominantly colonized with the nonpathogenic *E. dispar*, whereas in Taiwan and Japan extraintestinal disease due to *E. histolytica* is seen more commonly in HIV-infected individuals [9]. In a South African study, the seroprevalence of *E. histolytica* in HIV-infected individuals was three times that of HIV-negative patients, and those with a CD4 count lower than 200 were more likely to have a positive test [10]. Other risk factors for amoebiasis include young age, corticosteroid therapy, pregnancy, malnutrition, and institutionalized individuals [11].

LIFE CYCLE

There are many species within the genus Entamoeba: *Histolytica*, *Dispar*, *Moshkovskii*, *Coli*, *Hartmanni*, and *Polecki* are found in the intestinal tract [12]. Of these, *E. dispar* and *Moshkovskii* are identical to *E. histolytica*, but are nonpathogenic in humans [13]. *E. histolytica* is found in excreted cysts in contaminated water, food, and contaminated hands of food handlers. These cysts are able to survive in the environment for up to 2 months. Following ingestion of contaminated material, or during oroanal sexual practices (*anilingus*) [14], the cysts are released in the terminal ileum or colon where they become trophozoites. These either encyst to be excreted in feces, invade underlying mucosa to cause intestinal disease, or reach the blood vessels, for spread to extraintestinal tissues, e.g., liver, lung, brain, and others. The life cycle is depicted schematically in Fig. 18.1.

PATHOPHYSIOLOGY

The inciting event in disease causation in amoebiasis is adherence of *E. histolytica* trophozoites to the colonic epithelium (Fig. 18.2). The mechanisms determining whether there is encystation or invasion are incompletely understood; however, a protein called galactose/*N*-acetylgalactosamine (GAL/GalNAc)-specific lectin is implicated in mediating the adherence of the trophozoites to the colonic mucosa, thereby causing invasive disease [11]. *E. histolytica* as the name suggests causes lysis of tissue due to secreting proteinases, which digest extracellular proteins and thereby facilitate invasion into submucosa. In amoebic colitis, lateral extension of tissue necrosis through the submucosa layer results in the classic flask-shaped ulcer of amoebiasis. Cell damage can also be caused by the induction of apoptosis and an inflammatory response with the release of mediators such as interleukins and cyclo-oxygenases [11]. In amoebic liver abscess, one sees a well-circumscribed area of dead hepatocytes and debris surrounded by a rim of connective tissue

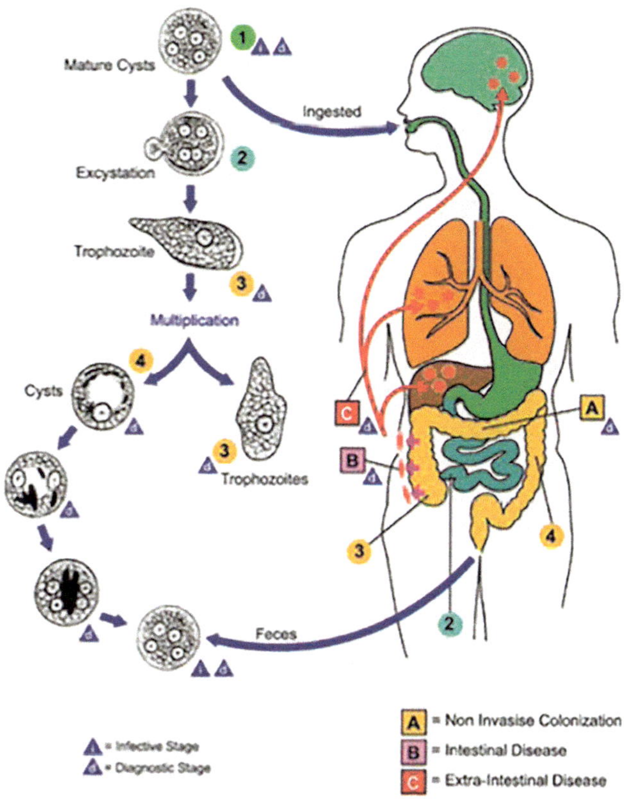

FIG. 18.1

Life cycle of amoebiasis.

Say hello to our little friends. Available from: https://sayhellotooutlittlefriends.files.wordpress.com/2012/03/dia-2.jpg [29 August 2015]

containing a few inflammatory cells and trophozoites. The fluid aspirated from an amoebic abscess contains sterile pus mixed with debris and blood, giving it its characteristic reddish brown color referred to as "anchovy paste."

CLINICAL FEATURES

The majority of infections are asymptomatic and cleared. 10%–20% of patients, however, present with dysentery or manifestations of extraintestinal disease. The latent period from infection to the development of symptoms varies from days to months, but is typically one to 4 weeks. Dysenteric symptoms are the commonest presentation with amoebic liver abscess (ALA) accounting for the majority of those with bloodborne spread.

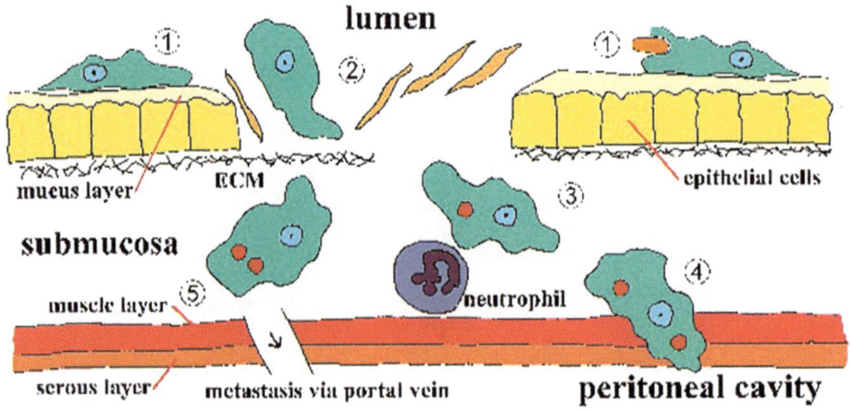

1. Trophozoites adhere to and colonize intestinal mucosa.
2. Breakdown of mucosal barrier with invasive disease.
3. Lateral and downward extension of trophozoites into submucosal layer causing a flask-shaped ulcer.
4. Penetration of muscle layer predisposing to intestinal perforation.
5. Trophozoites gain entry into the circulatory system for haematogenous spread.

FIG. 18.2

Pathogenesis of invasive amoebiasis.

Intestinal protozoa. Available from: http://www.tulane.edu/~wiser/protozoology/notes/eh_invade.html#one [29 August 2015]

GASTROINTESTINAL

In the gastrointestinal tract, asymptomatic carriage with passage of cysts in the stool is common. Patients with symptoms present with abdominal pain and diarrhea, and may later develop dysentery with a characteristic watery, bloody, and mucoid stool. Fever is an uncommon symptom in uncomplicated colitis (30% of individuals). The duration of increased stool frequency is variable and may be a self-limiting episode; however, when the stool frequency is florid and progressive and there is a systemic inflammatory response, severe disease is present and merits inpatient care and antimicrobial therapy. Occasionally, some individuals develop a fulminant colitis with high-grade fever, severe abdominal pain, peritonism, and profuse watery diarrhea [15]. Toxic megacolon complicates amoebic colitis in <0.5% of patients; nonetheless, these patients are at high risk of perforation (>75%) [16]. This risk is usually associated with states of immune compromise, for example, diabetes mellitus and the use of steroids [17]. Mortality is 40% higher than in those without fulminant colitis [15]. Luvuno et al., using angiography studies, described the pathological events in fulminant colitis. They showed that there is vascular occlusion with resultant transmural ischemia that incites a peritoneal reaction [18]. In some instances, however, the omentum wraps the area and contains or prevents imminent perforation. This process is often segmental and the intervening sections of bowel may be relatively normal producing a segmental colonic dilatation, although not as characteristic as that seen in ulcerative colitis.

Understanding of this process in the 1980s and 1990s led to the change in surgical management from a resection policy to one of prograde colonic lavage through the ileum, leaving the omentally contained ischemic areas to heal in situ [19]. This adaptation led to reduced mortality from 50% to 15%. The survivors often had resultant strictures, which could be resolved later when the acute inflammatory response had subsided and their nutritional status restored. This would involve resectional surgery tailored to the strictures [20]. Less commonly, with chronic amoebiasis, there may be the development of an inflammatory mass lesion or amoeboma that can mimic colonic cancer [21].

EXTRAINTESTINAL

Extraintestinal manifestations of amoebiasis represent invasive disease of which the most common presentation is amoebic liver abscess (ALA) [22]. ALA tends to have a male predominance. Compared to fulminant colitis, the mortality from ALA has decreased to 1%–3% [11] as a result of improved diagnosis and antimicrobials. ALA occurs as a result of hematogenous spread of amoebic trophozoites from invasive colonic disease via the portal veins to the liver. The majority of patients (95%) present within 5 months of leaving an endemic area, although this period may be as long as many years afterwards. Although multiple abscesses may occur (16%), the commonest presentation is that of a solitary abscess typically in the right lobe of the liver. Patients typically present with fever, right upper quadrant pain, and tenderness. The pain tends to be worse with breathing, coughing, and when lying on the right side. ALA is an important differential diagnosis in pyrexia of unknown origin, especially with a compatible travel history. Symptoms usually occur within a week or two, but if chronic, patients may in addition present with anorexia and loss of weight. These symptoms may or may not be associated with dysentery or the episode of dysentery may be remote. While jaundice is an unusual manifestation, the biochemical features commonly associated with ALA are leukocytosis and a raised alkaline phosphatase [23]. If the abscess is in the left lobe, the patient may present with epigastric discomfort and tenderness. In rare cases, an abscess may rupture into the peritoneum, intra-abdominal cavity, or pericardium [24]. The main differential diagnosis is that of pyogenic liver abscess, which has the same clinical features. Other differential diagnoses are a necrotic hepatoma and an infected echinococcal cyst.

PULMONARY

Pleuropulmonary complications occur in 20%–30% of patients, with an associated mortality of 6%. Atelectasis of the right lung and a sympathetic pleural effusion are common. Less common, but more serious is rupture of a large abscess into the pleural space or pericardium. In the former, patients present with cough, dyspnea, and pleuritic chest pain, whereas in the latter, patients are acutely ill with sudden hemodynamic compromise, with signs of pericarditis including cardiac tamponade. The mortality with rupture into the pericardium is as high as 30% [11]. A hepatobronchial fistula is a rare complication and when it occurs patients may expectorate copious amounts of anchovy paste-like material swarming with amoebae.

BRAIN

Brain abscesses due to *E. histolytica* are rare. They are most frequently seen in patients who were treated for ALA. The patients are usually young males who present with symptoms and signs of central nervous system involvement. Common presentations include headache, vomiting, and altered mental status.

The clinical findings include meningism, a facial nerve palsy, motor weakness, and seizures. Although the cerebrospinal fluid analysis is abnormal, there are no characteristic abnormalities [25].

OTHER PRESENTATIONS

Case reports have described amoebiasis presenting with rectovaginal [26] and cervico-uterine fistulae [27]. In addition, *E. histolytica* has been cultured in the knee [28] and skin [12].

DIAGNOSIS

There are several methods by which the diagnosis can be made each with advantages and disadvantages. The mainstay for the diagnosis of amoebic colitis and liver abscess is stool examination and serology, respectively. Identification of the cysts or trophozoites is key to making a diagnosis. The diagnostic modalities available are microscopy, antibody and antigen tests, culture, and molecular techniques such as polymerase chain reaction (PCR) [29]. Stool microscopy is particularly useful in resource-poor settings because it can be used for diagnosis and to differentiate *E. histolytica* from *E. dispar* by the presence of hematophagous entamoeba in the former, although this finding is not 100% specific [30]. The disadvantages of stool microscopy are the low sensitivity and specificity, the requirement of at least three fresh or frozen stool samples for analysis, the need for pathological expertise, and the fact that antimicrobials may interfere with the accuracy of the test. In symptomatic infection, serum antibodies are detectable in 70%–90% of cases [12]. Antibody tests that have been used include the amoebic gel diffusion test, indirect hemagluttin assays (IHA), enzyme-linked immunosorbent tests (ELISA), and others. The amoebic gel diffusion test was first used in the clinical setting in 1965 [31]. This test was considered highly positive when precipitation occurred within the first 20h, positive if within 40h, and negative if no precipitation had occurred at 48h. This test has largely been supplanted by the ELISA test, which is more quantitative, cost effective, and has a sensitivity between 90% and 100% and specificity of 94.8%, especially in ALA [32]. The disadvantage of ELISA is that it cannot differentiate between past and present exposure, particularly in endemic areas where patients have had multiple exposures. Antigen detection tests have been developed for diagnosis and are sensitive, specific, and easy to use. More importantly, they are able to distinguish between *E. histolytica* and *E. dispar*. Culture of *E. histolytica* and zymodene analysis has the advantage of being able to distinguish *E. histolytica* from *E. dispar*; however, this is time consuming, labor intensive, and the results are not rapidly available; therefore, they are not recommended for routine clinical practice. To overcome the diagnostic challenges of microscopic and culture identification, molecular biology techniques were developed; polymerase chain reaction (PCR) has the advantage of improved sensitivity (superior to antigen testing) and specificity, and simplicity of performance and together with ELISA, are the tests most recommended to make a diagnosis. It can be used in stool samples, tissue, and liver aspirate. The PCR tests in use target several amoebic genes including the small-subunit rRNA gene (18SrDNA), serine-rich *E. histolytica* protein (SREPH) gene [33], cysteine proteinases gene and actin genes [34]. The main drawback of PCR particularly in resource-poor settings is the cost. These diagnostic modalities will be discussed in relation to the two most common clinical presentations, i.e., dysentery and amoebic liver abscess.

AMOEBIC COLITIS

Stool collection and the demonstration of cysts or trophozoites on a wet mount preparation, or fixed specimen is essential. A sigmoidoscopic or colonoscopic examination may help identify ulcers and scrapings of the ulcer on a slide can identify hematophagous entamoeba with ingested red blood cells. Biopsies of the lesion should be performed to identify the typical histological feature of flask-shaped ulcer and trophozoites (Fig. 18.3). There is an available *E. histolytica* stool antigen/ELISA test on the market, specific for *E. histolytica*, with a sensitivity of 87% and specificity of 90% [16]. In amoebic colitis, it is also generally positive; however, in fulminant cases where patients are acutely ill or immunosuppressed, the test may be negative even in the presence of florid disease confirmed on stool microscopy. PCR techniques can also detect *E. histolytica* in stool specimens [35] and are significantly more sensitive than microscopy or culture, with 100% specificity for *E. histolytica* [36]. PCR is about 100 times more sensitive than fecal antigen tests [12].

AMOEBIC LIVER ABSCESS

The diagnosis is suspected on the basis of tenderness in the right upper quadrant and a systemic inflammatory response. Imaging of the liver and amoebic serology is the mainstays of the diagnosis. Ultrasound is the commonest initial diagnostic test recommended but a CT or MRI scan, which are more sensitive, may be performed. The ultrasonographic features are a solitary, cystic hypoechoic mass, typically in the right lobe of the liver. There may, however, be multiple lesions in which case a pyogenic abscess should be considered. On CT scanning, the lesion appears as a low-intensity mass with an enhancing outer rim [16]. In a patient who is well, repeat imaging is not warranted because the radiological features may persist for 2 years or longer. In a study by Simjee et al., the lesions persisted for up to 300 days in some cases [37].

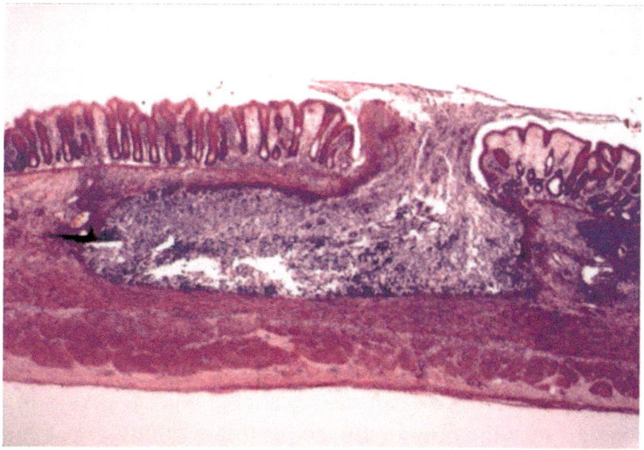

FIG. 18.3

Histopathology of a typical flask-shaped ulcer of intestinal amoebiasis.

Available from: http://www.pixnio.com [06 August 2018].

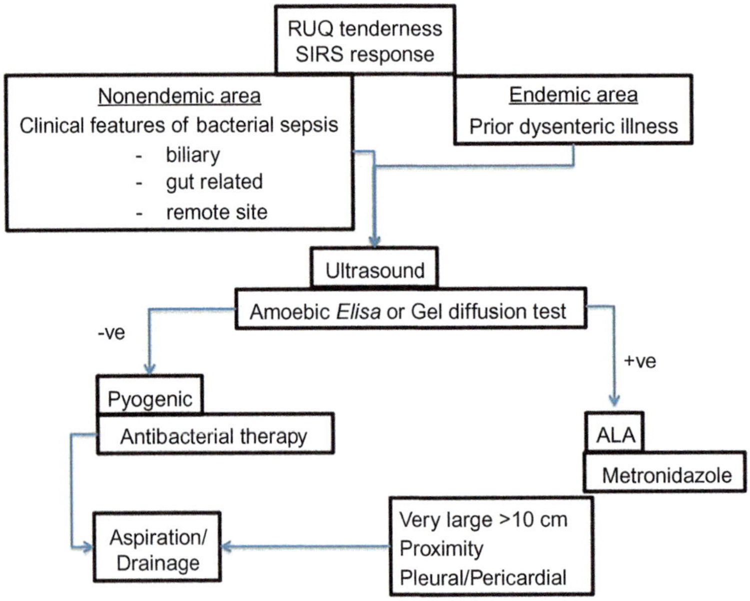

FIG. 18.4

Approach to ALA and indications for aspiration.

A chest X-ray is useful in ALA because half the patients will have elevation of the diaphragm. Almost all patients with ALA produce antibodies; however, these may be undetectable in the first week. Amoebic serology if positive confirms the diagnosis but if negative should prompt aspiration of the abscess to exclude a pyogenic abscess (Fig. 18.4). This is the main reason to aspirate an ALA as they are essentially areas of sterile necrosis. Some authors have advocated aspiration based on their size and proximity to the diaphragm or pericardium for fear of rupture into these cavities that may likely result in a fatal outcome. Other reasons to aspirate are for symptom relief or failure to improve clinically despite appropriate antimicrobial therapy, and to shorten hospital stay (Fig. 18.4). In addition to amoebic serology from blood samples, aspirated material can be sent for antigen or PCR testing to establish a diagnosis.

TREATMENT

All patients with *E. histolytica* infection should be treated, whether they are symptomatic or not (Fig. 18.5). This is done to prevent invasive disease. Patients with *E. dispar* or *E. moshkovskii* do not warrant therapy. Metronidazole is the drug of choice [38]; it is highly effective (>90% cure) administered orally and should only be given systemically if the patient is unable to tolerate oral intake. The recommended dose is 500–750 mg, three times a day, for 7–10 days; shorter durations of therapy are generally not recommended [39]. The elimination of intraluminal cysts agents using paromomycin is advised. Other agents such as tinidazole, ornidazole, and nitazoxanide are equally effective in treating *E. histolytica* [37]. These agents however are not available in resource-poor settings.

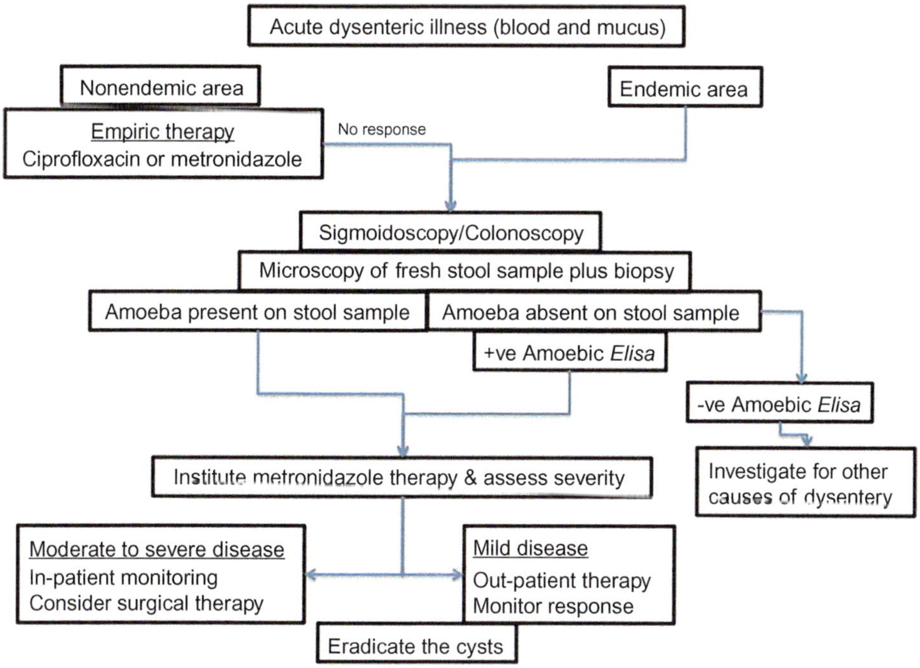

FIG. 18.5

Approach to diagnosis and management of dysentery.

PROGNOSIS AND OUTCOME

The prognosis of amoebiasis is good in general if the diagnosis is made timeously and therapy instituted promptly. The widespread usage of metronidazole as first-line therapy in endemic areas has led to a general decline in the incidence of amoebiasis, as such other diseases are now superseding amoebiasis as the cause of the dysentery. In addition, more severe forms of the disease, i.e., fulminant colitis and very large ALA that has ruptured into the pericardium, pleural space, and peritoneal cavity where mortality occurs are seen much less frequently.

VACCINE DEVELOPMENT

The role of innate and adaptive immunity in the pathogenesis of amoebiasis has not been well studied; however, it is known that mucosal IgA is secreted in invasive disease [3], making a vaccine feasible. Notwithstanding despite the presence of antibodies, recurrent infection is possible, suggesting that partial immunity is conferred following infection. Nonhuman primates, e.g., baboons have been used for vaccine development [40], but these studies are in their infancy. More recently, candidate vaccines have been tested in gerbils [41] and mice [42], with varying success. Human vaccines against amoebic proteins are further down the pipeline, but would be useful in endemic areas or for travelers to endemic areas.

CONCLUSIONS

Diagnosis and management of amoebiasis has significantly improved outcomes despite the HIV epidemic in sub-Saharan Africa. Clinical and scientific research has contributed to solutions to a variety of clinical challenges in the diagnosis, management, and complications of this disease. The condition however still accounts for significant morbidity and mortality in complicated cases.

TAKE-HOME MESSAGES

- Only *E. histolytica* is pathogenic in humans.
- It is spread via the feco-oral route.
- Infection is typically asymptomatic; when symptomatic patients present with dysentery or ALA.
- Diagnosis in resource-poor settings by stool microscopy is still a valuable diagnostic tool.
- Metronidazole is the drug of choice for invasive disease and paromomycin for cyst eradication.
- ALA requires aspiration.

REFERENCES

[1] Bercu TE, Petri WA, Behm JW. Amebic colitis: new insights into pathogenesis and treatment. Curr Gastroenterol Rep 2007;9(5):429–33.

[2] Caballero-Salcedo A, et al. Seroepidemiology of amebiasis in Mexico. Am J Trop Med Hyg 1994;50(4):412–9.

[3] Haque R, Petri Jr. WA. Diagnosis of amebiasis in Bangladesh. Arch Med Res 2006;37(2):273–6.

[4] Sargeaunt PG. Amoebiasis. J R Soc Med 1982;75(11):920–1.

[5] Gathiram V, Jackson TF. Frequency distribution of *Entamoeba histolytica* zymodemes in a rural South African population. Lancet 1985;1(8431):719–21.

[6] Jackson T, et al. A comparison of cross-sectional and longitudinal seroepidemiological assessments of entamoeba-infected populations in South Africa. Arch Med Res 2000;31(4 Suppl):S36–7.

[7] Stauffer W, Abd-Alla M, Ravdin JI. Prevalence and incidence of *Entamoeba histolytica* infection in South Africa and Egypt. Arch Med Res 2006;37(2):266–9.

[8] Moran P, et al. Infection by human immunodeficiency virus-1 is not a risk factor for amebiasis. Am J Trop Med Hyg 2005;73(2):296–300.

[9] Hung CC, et al. Increased risk for *Entamoeba histolytica* infection and invasive amebiasis in HIV seropositive men who have sex with men in Taiwan. PLoS Negl Trop Dis 2008;2(2):e175.

[10] Samie A, et al. Seroprevalence of *Entamoeba histolytica* in the context of HIV and AIDS: the case of Vhembe district, in South Africa's Limpopo province. Ann Trop Med Parasitol 2010;104(1):55–63.

[11] Stanley Jr. SL. Amoebiasis. Lancet 2003;361(9362):1025–34.

[12] Fotedar R, et al. Laboratory diagnostic techniques for Entamoeba species. Clin Microbiol Rev 2007;20(3):511–32, table of contents.

[13] Clark CG, Diamond LS. Ribosomal RNA genes of 'pathogenic' and 'nonpathogenic' *Entamoeba histolytica* are distinct. Mol Biochem Parasitol 1991;49(2):297–302.

[14] Salit IE, et al. A possible cluster of sexually transmitted *Entamoeba histolytica*: genetic analysis of a highly virulent strain. Clin Infect Dis 2009;49(3):346–53.

[15] Takahashi T, et al. Fulminant amebic colitis: analysis of 55 cases. Dis Colon Rectum 1997;40(11):1362–7.

[16] Petri Jr. WA, Singh U. Diagnosis and management of amebiasis. Clin Infect Dis 1999;29(5):1117–25.

[17] Chaturvedi R, Gupte PA, Joshi AS. Fulminant amoebic colitis: a clinicopathological study of 30 cases. Postgrad Med J 2015;91(1074):200–5.

[18] Luvuno FM, Mtshali Z, Baker LW. Vascular occlusion in the pathogenesis of complicated amoebic colitis: evidence for an hypothesis. Br J Surg 1985;72(2):123–7.

[19] Luvuno FM. Role of intraoperative prograde colonic lavage and a decompressive loop ileostomy in the management of transmural amoebic colitis. Br J Surg 1990;77(2):156–9.

[20] Singh B, Moodley J, Ramdial PK. Fulminant amoebic colitis: a favorable outcome. Int Surg 2001;86(2):77–81.

[21] Misra SP, Misra V, Dwivedi M. Ileocecal masses in patients with amebic liver abscess: etiology and management. World J Gastroenterol 2006;12(12):1933–6.

[22] Rao S, et al. Hepatic amebiasis: a reminder of the complications. Curr Opin Pediatr 2009;21(1):145–9.

[23] Maltz G, Knauer CM. Amebic liver abscess: a 15-year experience. Am J Gastroenterol 1991;86(6):704–10.

[24] Chuah SK, et al. The prognostic factors of severe amebic liver abscess: a retrospective study of 125 cases. Am J Trop Med Hyg 1992;46(4):398–402.

[25] Petri WA, Haque R. *Entamoeba histolytica* brain abscess. Handb Clin Neurol 2013;114:147–52.

[26] Fadiran OA, et al. Amoebic recto-vaginal fistula—a case report and review of literature. Cent Afr J Med 1993;39(8):172–5.

[27] Angulo GB, Sanchez DR. Cervico-uterine amebiasis. Characteristics of 15 cases and review of the literature. Rev Latinoam Microbiol 1980;22(2):105–8.

[28] Than S, et al. Isolation of *Entamoeba histolytica* from arthritic knee joint. Trop Geogr Med 1992;44(4):355–8.

[29] Tanyuksel M, Petri Jr. WA. Laboratory diagnosis of amebiasis. Clin Microbiol Rev 2003;16(4):713–29.

[30] Haque R, et al. Rapid diagnosis of Entamoeba infection by using Entamoeba and *Entamoeba histolytica* stool antigen detection kits. J Clin Microbiol 1995;33(10):2558–61.

[31] Powell SJ, et al. Amoebic gel-diffusion precipitin-test. Clinical evaluation in amoebic liver abscess. Lancet 1965;2(7413):602–3.

[32] Hira PR, et al. Invasive amebiasis: challenges in diagnosis in a non-endemic country (Kuwait). Am J Trop Med Hyg 2001;65(4):341–5.

[33] Stanley Jr. SL, et al. Cloning and expression of a membrane antigen of *Entamoeba histolytica* possessing multiple tandem repeats. Proc Natl Acad Sci U S A 1990;87(13):4976–80.

[34] Freitas MA, et al. A single step duplex PCR to distinguish *Entamoeba histolytica* from Entamoeba dispar. Parasitology 2004;128(Pt 6):625–8.

[35] Roy S, et al. Real-time-PCR assay for diagnosis of *Entamoeba histolytica* infection. J Clin Microbiol 2005;43(5):2168–72.

[36] Blessmann J, et al. Real-time PCR for detection and differentiation of *Entamoeba histolytica* and Entamoeba dispar in fecal samples. J Clin Microbiol 2002;40(12):4413–7.

[37] Simjee AE, et al. Serial ultrasound in amoebic liver abscess. Clin Radiol 1985;36(1):61–8.

[38] Gonzales ML, Dans LF, Martinez EG. Antiamoebic drugs for treating amoebic colitis. Cochrane Database Syst Rev 2009;15(2):Cd006085.

[39] Singh G, Kumar S. Short course of single daily dosage treatment with tinidazole and metronidazole in intestinal amoebiasis: a comparative study. Curr Med Res Opin 1977;5(2):157–60.

[40] Abd Alla MD, et al. Adherence-inhibitory intestinal immunoglobulin a antibody response in baboons elicited by use of a synthetic intranasal lectin-based amebiasis subunit vaccine. Infect Immun 2007;75(8):3812–22.

[41] Ivory CP, Keller K, Chadee K. CpG-oligodeoxynucleotide is a potent adjuvant with an *Entamoeba histolytica* Gal-inhibitable lectin vaccine against amoebic liver abscess in gerbils. Infect Immun 2006;74(1):528–36.

[42] Houpt E, et al. Prevention of intestinal amebiasis by vaccination with the *Entamoeba histolytica* Gal/GalNac lectin. Vaccine 2004;22(5–6):611–7.

CHAPTER

INTESTINAL SCHISTOSOMIASIS IN AFRICA

19

Suliman Salih Fedail

Chairman of Fedail Hospital, Consultant Physician and Gastroenterologist, Khartoum, Sudan

INTRODUCTION

Schistosomiasis is caused by infection with a helminthic blood parasite of the genus schistosome. It inhabits human vasculature. It is an old disease in Africa found in the Egyptian mummies. Approximately 207 million people in the world are infected with schistosomiasis; of these, 90% live in sub-Saharan Africa.

The disease is most prevalent in Nigeria and Tanzania. There is a recent increase in its prevalence in Senegal due to construction of Diama Dam and in Malawi due to overfishing in lake Malawi as fish consume the snails, the intermediate host of schistosomiasis. There are also successes stories as it has been eradicated from Tunisia and Morocco. Prevalence in Egypt had dropped from 40% to <3% between 1957 and 2006.

Schistosomiasis is among the list of neglected tropical diseases. Schistosomiasis rank second to malaria as the most devastating disease in Africa. Almost 300,00 people die annually from schistosomiasis in Africa.

The most common way of getting schistosomiasis is by wading, swimming or bathing in lakes, ponds, irrigation canals, and other bodies of fresh water that is infested with snails, which are the natural reservoir of schistosoma parasites (Fig. 19.1).

In Africa, Schistosomiasis is due mainly to infection with schistosomiasis mansoni, which causes intestinal schistosomiasis and schistosoma hemtoburn, which causes urinary schistosomiasis.

In this chapter, I will discuss the burden of schistosoma mansoni (intestinal schistosomiasis) in Africa, clinical picture, diagnosis, treatment, and prevention.

LIFE CYCLE

The life cycle of schistosomiasis is also known as Bilharsiasis after the German doctor Theodore Bilharz who was the first to indentify the parasite in Egypt in 1852. It requires both an intermediate host (snail) and a final host (human). Schistosomes are unsegmented leaf-like organisms. The adult worm is approximately 1–2 cm long, with a cylindrical body, two terminal suckers, digestive tract, and reproductive organs. The male schistosoma is larger and folded forming a gynecophoric canal or "Schist" where the slender female resides.

Digestive Diseases in Sub-Saharan Africa. https://doi.org/10.1016/B978-0-12-815677-3.00019-0

FIG. 19.1

Children playing in canal infested with schistosmiasis.

The adult worms digest erythrocytes. Schistosomes have no anus, so excretes waste by regurgitating it in the blood stream. These expelled products are utilized for diagnostic tests.

The precise nature of biochemical and reproductive exchange between the two sexes is unknown. The female produces hundreds to thousands of eggs per day. When humans discharge schistosomes eggs in water, each egg releases a ciliated larva, the miracidium, which seeks a snail, the intermediate host.

The miracidium enters the snail to undergo asexual reproduction, within approximately one month, it undergoes a series of developmental changes until the final larva stage of the parasite, the cercaria is produced and is ejected by the snails into the water. Cercaria are the form of the parasite infectious to human beings. The intermediate host for schistosoma mansoni is aquatic fresh water snail of Biomphalaria species. Snails once infected are infected for life, which is usually about 12 months, act as a photocopier producing thousands of cercariae daily. The cercariae are small, less than 1 mm long, with bifurcated tail and are either male or female, they survive up to 48 h in water, they are attracted to humans by certain chemicals found in human skins. The cercaria once attached to the skin secretes enzymes that enable it to enter into the skin, inside the human skin, the cercaria sheds its forked tail and transforms into preadult schistoumulum. Schistosomulae are tailless and worm like in appearance, migrate from skin into blood and lymph vessels, and are carried to the heart; they eventually reach the liver where they live temporarily for 1–4 weeks, in the liver they mature into adult worms, the mature worms pair in the liver, the longer female residing in the gynecophoric channel of the shorter male, they remain in a constant state of copulation. The adult worms migrate against portal blood flow to the inferior mesenteric veins and superior hemorrhoidal veins and deposit their eggs mainly in the rectum sigmoid and descending colon. Ova are usually found in the loose submucosa of the large intestine.

What guides adult schistosomes to its final destination in the intestines is unknown. The adult worms remain in these blood vessels for life, they adhere to the vessel wall by their suckers. Worms usually survive for 5–10 years but survival up to 30 years was reported, the adult schistosoma does not replicate within the host. In 1 3 months time after reaching their final residence, the female worm produces eggs, which travels from the vessels through host tissue to the lumen of intestine or carried by blood to other sites in the body.

PATHOGENESIS

The adult worms do little damage to host and cause no symptoms unless they are lost and settle in an unusual location, e.g., brain or spinal cord. The major pathogenesis is caused by schistosomal eggs when they are trapped in tissues, which leads to fibrosis and scarring of affected tissues. The egg induces granuloma formation in a delayed type of hypersensitivity reaction. In the large intestines, eggs lead to ulceration and scarring, which leads to blood in stool. Eggs, which are carried by portal blood flow to the liver, wedge in the presinusidal radicals of the portal vein; this elicits a fibrotic reaction that eventually blocks the venous blood flow of the liver. The characteristic pathology is periportal fibrosis described by Symmers as pipe-stem fibrosis; this occurs usually 5–15 years after the onset of the infection, and it is the most important, most dreaded, and costly complication of schistosomiasis. The presinusidal blockage leads to the development of severe portal hypertension and formation of porto-systemic collaterals, varices, at the esophagus, stomach, and other sites; this leads to recurrent gastrointestinal bleeding. In a field study, a village in the endemic area in Gezira central Sudan, 54% of people living in the village were found to have esophageal varices. The hepatocytes are not affected as in liver cirrhosis because liver cell perfusion is preserved, the nodular architecture of the liver is maintained, and nodular regenerative hyperplasia does not occur. There is no evidence to link schistosomiasis to colonic cancer or liver cancer.

CLINICAL FEATURES

Most acute infections in endemic areas are asymptomatic.

SWIMMER'S HCH

This is a skin reaction, which occurs at the site of cercaria entry typically on the lower legs and feet. It may occur up to 2–3 days after penetration of cercaria into the skin. It causes a transient pruritic maculopapular rash. This is usually a self-limiting condition, symptoms rarely last more than a week. The condition is particularly seen when humans are exposed to avian schistosomes. Prior sensitization to cercariae results in a more rapid and severe symptom.

KATAYAMA FEVER

It occurs in areas of high transmission rates. History of contact with contaminated water 2–8 weeks earlier is usual. It is thought to be a systemic hypersensitivity reaction against the migrating parasites egg production and release of egg antigens. Katayama fever is a serum sickness like syndrome with fever, headache, mylagia, generalized lymphadenopathy, and hepatosplenomegaly. The symptoms usually resolve spontaneously over a period of few weeks, rarely patients develop patchy infiltrate on chest X-rays. Aseptic meningitis and death have been reported. The diagnosis of Katayama fever relies on appropriate history of exposure in addition to the aforementioned clinical features. Eggs are rarely found, most patients have positive serological tests, marked eosinophilia is a very important surrogate marker. This syndrome is rarely seen in resident of endemic areas possibly due to in utero priming of lymphocytes of babies born to mother with helminthic infections.

CHRONIC INFECTION OF THE LARGE INTESTINE

It presents with intermittent bloody diarrhea with tenesmus and colicky lower abdominal pain. Chronic fatigue from anemia is often complained of. On examination, the patient may be pale, lower abdominal tenderness is frequently elicited. Complications of longstanding intestinal schistosomiasis include colonic ulcers, colonic strictures, and colonic polyps. Colonic polyps may present as inflammatory masses, which may even mimic cancer. Rarely colonic polyposis results in protein—loosing enteropathy. The severity of symptoms and complications are related to the intensity of infection.

HEPATIC SCHISTOSOMIASIS

The patient develops hematemesis from esophageal or gastric varices or very rarely from ectopic varices. The hematemesis is usually of sudden onset without prodromal symptoms, rarely it is preceded by dizziness, fatigability, and epigastric discomfort. Melaena can be the first presentation but usually follows hematemesis. The patient is pale, tachycardic, and hypotensive. The liver is shrunken, left lobe may be enlarged, the spleen is usually palpable but after severe bleeding becomes palpable only after resuscitation. Ascites is very rare unless the bleeding is massive due to hepatocellular ischemia. Jaundice and features of chronic liver disease or hepatic encephalopathy are rare, as the hepatocytes are normal in these patients. If there are signs of hepatic decompensation, one should suspect coinfection with hepatitis viruses B or C. Some patients may present with symptoms of anemia due to chronic blood loss from varices, portal hypertension gastropathy, or hypersplenism. Dragging left hypochondrial pain from gross splenomegaly is reported by some patients. Hepatorenal syndrome and spontaneous bacterial peritonitis are rarely seen in pure schistosomiasis.

The differential diagnosis includes visceral leishmaniasis, tropical splenomegaly syndrome, blood disorders, and liver cirrhosis.

PULMONARY SCHISTOSOMIASIS

Eggs can pass to the lungs via portosystemic anatomsis causing granulomatous pulmonary arteritis leading to pulmonary hypertension. Dyspnea is the principal symptom. The pulmonary changes are irreversible.

NEUROSCHISTOSOMIASIS

This may result from adult worms migrating aberrantly to the brain and spinal cord. Schistosoma eggs may embolize to the spinal cord and brain. Brain involvement may lead to focal epilepsy, affection of the spinal cord may lead to transverse myelitis or subacute myleoradialopathy and may result in permanent paralysis.

PEDIATRIC SCHISTOSOMIASIS

Schistosoma infection during childhood causes substantial growth retardation, anemia, and affects cognitive development. This is in addition to the known chronic complications of schistosomiasis; these are reported in children as young as 6 years old.

SCHISTOSOMIASIS AND OTHER INFECTIONS

Chronic Salmonella infection

The Salmonella bacteria colonizes the adult worm that act as a reservoir of infection. Recurrent bacteriemia due to Salmonella may lead to recurrent attacks of enteric fever. Eradication of the adult worm is mandatory in order to get rid of the recurrent bacteremia.

Malaria

The effect of schistosomiasis on malaria infection is not known, most of the studies produced contradicting results.

Hepatitis viruses

The high prevalence hepatitis C virus infection in Egypt has been attributed to the wide use of parenteral tartar emetic preparations in the 1950s and 1960s to treat schistosomiasis. However, similar use of the drugs in Sudan did not result in hepatitis C virus epidemic. In Egypt, the prevalence of hepatitis C virus is around 20%, in Sudan it is only 2.5%.

Concomitant infection with hepatitis B and C results in a more severe liver disease and worse prognosis. We advise patients with periportal fibrosis to be vaccinated against hepatitis B.

Schistosomiasis mansoni and human immunodeficiency virus (HIV)

HIV patients infected with *S. mansoni* found to excrete fewer eggs in stools. This observation is compliable with hypothesis that the process of excretion of eggs requires an active inflammatory reaction. Females with genital schistosomiasis have three or four times increased risk of HIV infection possibly due to compromised physical barrier.

DIAGNOSIS OF SCHISTOSOMIASIS

MICROSCOPY

Schistosoma mansoni eggs can easily be detected in stool and identified by microscopy due to their size, shape, and their typical lateral spine. It is cheapest and the mostly widely used method to diagnose schistosomiasis in endemic areas. In field studies, the Kato-Katz concentration method is a rapid, simple, and inexpensive method to quantify the amount of eggs in stool and is recommended by World Health Organization (WHO). Molecular techniques to detect schistosome DNA in fecal specimens are sensitive but are not widely used. Schistosoma eggs may be demonstrated in tissue biopsy specimens from the intestine and liver (Fig. 19.2). Eggs may rarely be found in cerebrospinal fluid of patients with spinal cord involvement.

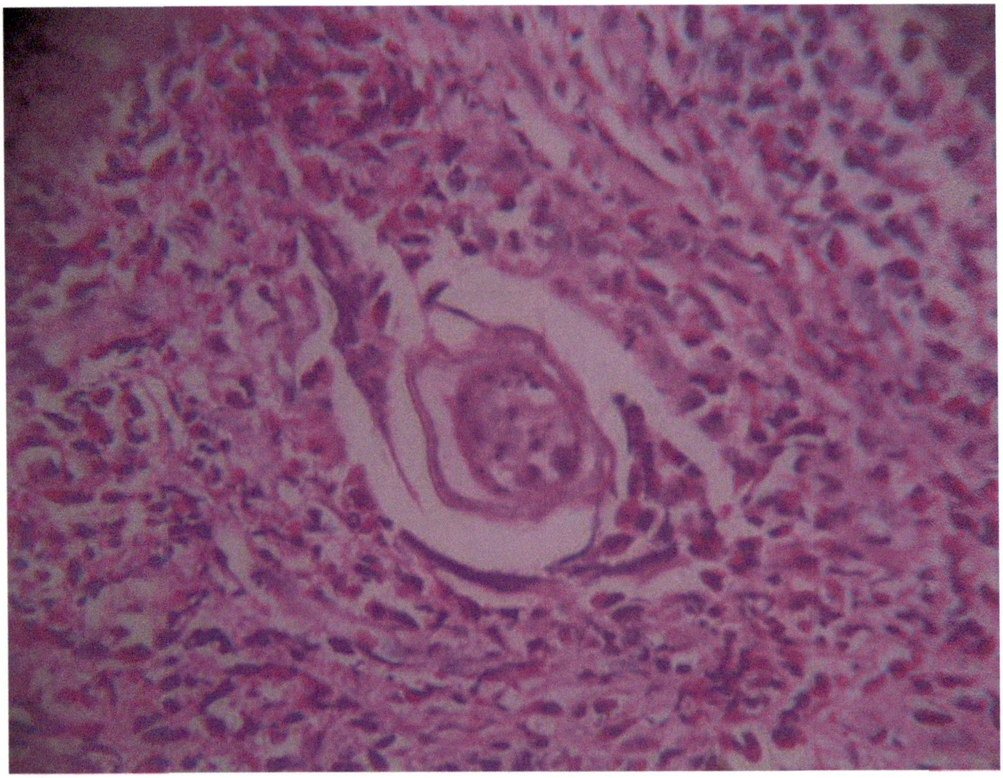

FIG. 19.2

Schistosoma granuloma in the liver.

SEROLOGY

Serologic tests are available that can detect antischistosomal antibodies in the serum, mostly ELISA based; however, none of them can distinguish between a past infection and a current active disease, antibody tests also cross react with other helminths. Serology is a more specific test for travelers who are not otherwise expected to have schistosomal antibodies. Seroconversion generally occurs within four to eight weeks after infection. A positive serologic test may be diagnostic in patients in whom there are no eggs in excreta such as those with Katayma Fever. Serology is also helpful in the difficult central nervous system schistosomiasis when eggs cannot be recovered. Serological tests are useful in field studies especially to detect whether infection re-emerged in a region after an apparently successful eradication program. Antigen-based detection tests have been developed, they enable on-site mapping of schistosoma mansoni infection without stool collection.

OTHER TESTS

Anemia may be found in chronic cases due to blood loss. Pancytopenia due to hypersplenism occurs. Routine white cells count may show eosinophilia in infected patients. Liver function tests are usually normal as the hepatocytes are well preserved, they only become abnormal after a recent massive gastrointestinal bleeding or a concomitant infection with hepatitis viruses. Ultrasound is very valuable in showing the classical periportal fibrosis (Fig. 19.3), which is easy to detect and can immediately confirm the diagnosis of hepatic of schistosomiasis as no other disease causes these typical changes. Computed tomography (CT) and magnetic resonance imaging (MRI) are expensive and mostly unavailable in endemic areas but also show classical features of periportal fibrosis. They add no more information than ultrasonography, which is cheap and available. CT and MRI might be useful in the evaluation of central nervous system disease. Recently, PET scans are used experimentally to detect adult parasites in vivo, but this is unlikely to be of use due to its high cost.

Gastroscopy will show esophageal, gastric, or duodenal varices. Sigmoidoscopy and colonoscopy show patchy erythema and shallow ulcers; sometimes, there is diffuse erythema, which looks like ulcerative colitis. They may show rectal varices.

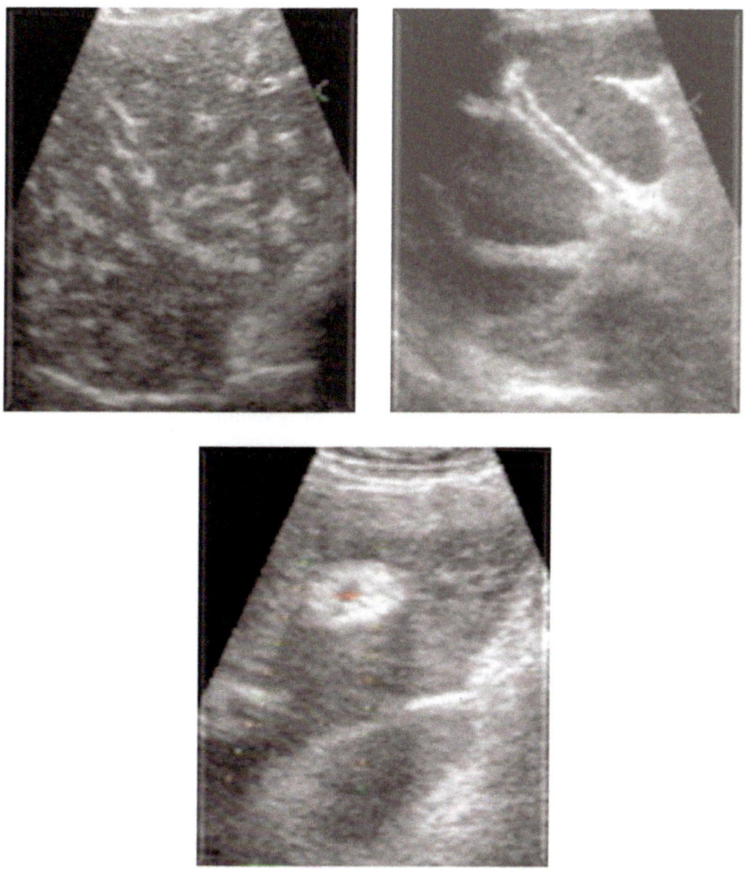

FIG. 19.3

Symmer pipe-stem fibrosis.

TREATMENT

SWIMMER'S TECH

Clears up in few days without treatment. Anti-itching creams, e.g., Calamine lotions might help. In severe cases, oral antihistamines will reduce itching.

KATAYMA FEVER

It is primarily treated with corticosteroids to suppress the hypersensitivity reaction, usually prednisolone tablets 30–45 mg daily for 10 days. Praziquantel is given initially and after 6 weeks to eradicate the mature worms at onset of the illness and the worms, which will mature later.

TREATMENT OF CHRONIC SCHISTOSOMIASIS

The drug therapy of schistosomiasis was first initiated with the use of tartar emetic (Potassium antimony tartarate) in 1918 by the British physician Dr. J.B. Christopherson in Khartoum Civil Hospital, Sudan. In the following years, other antimony salts were used. These drugs were effective but toxic and have to be given parenterally. Lately, a lot of oral treatments were used. Praizquantel, however, is now almost exclusively used in the treatment of schistosomiasis. Praziquantel is a pyrazino-isoquioline derivative, which is effective against all the major five species of human schistosomiasis; it is given orally at a single dose of 40 mg/kg body weight, use of higher dose did not prove more effective. It is marketed in tablets of 600 mg. The exact mechanism of its action in the worm is unknown, but it is thought to produce paralysis of the adult worm, resulting in its detachment from the vein wall. It is well tolerated and cause few side effects, mainly nausea, vomiting, malaise, and abdominal pain; most of these symptoms are thought to be caused by immune response to dying worms. It is safe for treatment in pregnancy, lactation, and in young children. It is even reported to reduce fibrotic changes in the liver. Praziquantel has little or no effect on schistosoma eggs or immature worms.

There is increasing concern about the development of resistance to praziquantel, but fortunately it has not been materialized. In Sudan it costs 40 US cents per tablet, this is mainly due to new manufacturers in Korea, China, Egypt, Sudan, Brazil, and other countries. This has led to its use also in control of schistosomiasis using mass chemotherapy. Other drugs like oxmniquine, which is effective against *S. mansoni* only, was used extensively in Brazil but has now been replaced by praziquantel. Artemisinin derivatives, which are the main drugs for the treatment of malaria, extracted from the leaves of a Chinese sweet worm wood, were shown to be effective against human schistosomes. There is a real concern about their use in treatment of schistosomiasis in areas where malaria is endemic for fear of inducing malaria resistance to artimisinuim derivates. The recent mapping of the schistosoma mansoni genome, which showed that it has seven pairs of autosomal chromosomes and one pair of sex chromosomes, might help in developing new drugs and methods of treatment for schistosomiasis.

TREATMENT OF CHRONIC COMPLICATIONS OF SCHISTOMIASIS

The chronic complications of intestinal schistosomiasis are the main cause of morbidity and mortality and reduction of productivity and the most expensive to treat. Bleeding esophageal varices is the leading presentation of hepatic schistosomiasis. Blood transfusion is required and it is always in short supply as almost everyone in the village has varices and cannot donate. Iron supplement is given regularly. The varices need to be properly treated in hospitals with endoscopy facilities to inject them with sclerosants or to band ligate them. Ethanolamine oleate is widely used for esophageal varices and N-butyl-2-cyanoacrylate is very effective in gastric varices. Endoscopic treatment of varices is very effective, in cases of hepatic schistosomiasis but the patient needs to come monthly initially till the varices are eradicated and then 1–2 yearly to check and treat new varices for life. Nonselective B-Blockers, mainly propranolol, is used as a primary prophylaxis in those with large varices who never bled and as secondary prophylaxis alone if there are no endoscopic facilities or in addition to endoscopic treatment where endoscopy is available till they are given for life. Tolerance to B-blocker is very good and few stop the treatment due to side effects. The response to B-blocker and endoscopic treatment is excellent as liver functions are well maintained and we have patients undergoing regular sclerotherapy and then band ligation for over 25 years. Sengstaken tubes are useful for temporary arrest of bleeding and for transfer of patients to endoscopy units. Inflation of the gastric balloon alone is enough. Recently introduced self-expandable metallic stents for treatment of esophageal variceal bleeding may have a great potential in developing countries as their introduction does not need expensive X-ray equipment.

Surgery is used only when there is hypersplenism to remove the spleen; it is no longer used to treat bleeding esophageal or gastric varices.

The treatment of colonic complications of colonic schistosomiasis is by giving praziquantel, small polyp dissolve with drug treatment, large polyps need endoscopic polypectomy.

Terlipressin is very effective but expensive, used only on the rare patient who continued to bleed after endoscopic therapy. Transjugular intrahepatic protosystemic shunt (TIPS) is very expensive and is used in very few cases, it is difficult to introduce the stent as liver is fibrotic and hard. Portal hypertension gastropathy is rarely seen and responds to B-blockers therapy, gastric vascular ectasia is also very rare and is treated with Argon plasma coagulation. Antibiotics are not needed in the management of acute bleeding in schistomiasis as bacterial infection is very rare, unlike the case in liver cirrhosis.

CONTROL

WHO is recommending population control of schistosomiasis based on treatment with praziquantel yearly indefinitely, as there is little natural immunity to reinfection to the exposed population. The WHO envisaged a gradual eradication of schistosomiasis. It endorsed in 1984 a strategy to control morbidity caused by schistosomiasis through preventive chemotherapy with praziquantel every 1–2 years, which can be given by village health workers or teachers. In 2012, WHO recommended that countries should aim beyond control of morbidity toward elimination of schistosomiasis. This is made possible by Merck Serono drug company who pledged to donate up to 250 million tablet of Praziquantel per year. The WHO aims to eradicate schistosomiasis from the world by 2025, which is very ambitious.

In endemic areas, people purchase praziquantel over the counter and use it. This definitely reduced the prevalence rate by 90% in some countries. This may be difficult to maintain indefinitely, a vaccine might solve the problem of prevention in the long term. However, till now, no clinical trials for human vaccine to schistosomiasis have shown to be effective. Transmission control using mulluscicides is no longer a practical option.

We should not forget that millions of people who are currently infected may be living for at least the next 50 years with periportal fibrosis and bleeding varices. These patients need good endoscopic facilities to treat them, as if bleeding varices are eradicated endoscopically, these people can lead a normal life span. Unfortunately, the WHO and major donor organization do not want to participate in this important management aspect. Eradicating varices will enable these people to go back to work earn money, build bathrooms and toilets in their homes, thereby preventing their children from going to canals and getting infected with schistosomiasis.

Proper planning and innovative management of water resource schemes for power generation and irrigation is very important. The building of Sennar dam in the Blue Nile in Sudan in 1925 introduced schistosomiasis for the first time in central Sudan and resulted in immense suffering of the population till today. More recently, the erection of Diama dam in Senegal in 1986 introduced schistosomiasis to Senegal and Mauritania; by 1994 virtually the whole population of Northern Senegal became infected with schistosomiasis.

Poverty, lack of risk awareness, insufficient public health facilities and lack of basic life amenities like safe water and sanitation, the absence of political will by politicians had contributed to high prevalence of the disease in Africa. Health education behavioral changes, provision of funds, and stable governments will accelerate the eradication of schistosomiasis in Africa.

The final solution to the schistomiasis problem is to eradicate poverty, which is its principal cause.

RECOMMENDED READING

[1] Collen DG, Bastindy AL, Evan Secuv W, King CH. Human schistosomiasis. Lancet 2014;383:2253–3364.

[2] Rokn MB, editor. Schistosomiasis. Rijeka: InTech; 2012.

[3] Gryseels B, Polman K, Clernix J, Kestens L. Human schistosomiasis. Lancet 2006;368:1106–18.

[4] Ross AGP, Bartely PB, Sleigh AC, et al. Schistosomiasis. N Eng J Med 2002;346:1212–20.

[5] Mahmound AAFM, editor. Schistosomiasis. In: Tropical diseases science and practice, vol. 3. Singapore: Imperial College Press; 2001.

[6] Mudawi HMY, Smith HM, Fletcher IA, Fedail SS. Prevalence and common genotypes of HCV infection in Sudanese patients with hepatosplenic schistosomiasis. J Med Virol 2007;79:1322–4.

[7] Katzn N, Coelho PM. Clinical therapy of schistosomiasis mansoni: the Brazilian contribution. Acta Tropica 2008;108:72–8.

[8] Abdul Gani R, Loutfy N, El Sahra A, Hassan A. Current chemotherapy arsenal for schistosomiasis mansoni: alternatives and challenges. Parasitol Res 2009;104:955–65.

[9] Fedail S. Oesophageal varices in Sudan. Gastrointest Endosc 2002;56(5):781–2.

CHAPTER

CHOLERA IN SSA

20

Isidor Segal

G.I.T. Unit, Chris Hani Baragwanath Hospital and University of Witwatersrand, Soweto, South Africa

INTRODUCTION

Cholera is an acute secretory diarrheal disease that can kill within hours if left untreated. It is caused by Vibrio cholera of the O1 or O139 serogroup. The latter serogroup (0139) has only been identified in sporadic cases and has never been identified outside Asia (WHO [1]). It is a Gram-negative bacterium that produces cholera toxin, an enterotoxin, which acts on the mucosal epithelium lining of the small intestine. In its most severe form, it results in a sudden onset of acute watery diarrhea that can lead to death by severe dehydration and kidney failure. It has a short incubation period—2 to 5 days—about 75% of people infected with cholera do not develop any symptoms.

Each year there are more than 1.3 million to 4.0 million deaths worldwide due to cholera and other waterborne diseases.

Safe oral cholera vaccines should be used in conjunction with improvements in water and sanitation to control cholera outbreaks and for prevention in areas known to be high risk for cholera.

Global strategy on cholera control with a target to reduce cholera deaths by 90% was launched in 2017 (WHO! Feb 2018).

Other strains of *V. cholerae* apart from O1 and O139 can cause mild diarrhea but do not develop into epidemics.

EPIDEMIOLOGY

Cholera is endemic in many developing countries. Southern and East African countries are in constant threat of cholera outbreaks. This is due to poor sanitation and is an example of government neglect [2]. In 2017 there were 109,442 cholera cases resulting in 1708 deaths in 12 countries in the Eastern and Southern African Region. Since the beginning of 2018, there have been more than 2009 cases and a further 22 deaths in seven countries—Angola, Kenya, Malawi, Mozambique, Somalia, Tanzania, and Zambia. In Zambia, more than 74 people died of Cholera since October 2017 [2]. Other countries severely affected by Cholera are South Sudan, Democratic Republic of Congo, Burundi, Rwanda, Uganda, and Zimbabwe.

High-risk areas are slums because of the absence or shortage of safe water and sufficient sanitation.

Digestive Diseases in Sub-Saharan Africa. https://doi.org/10.1016/B978-0-12-815677-3.00020-7

TRANSMISSION

Cholera is transmitted through ingestion of water contaminated with the cholera bacterium, usually from feces of other cholera patients. Feces is discharged into waterways, groundwater, or drinking water supplies. Foods washed in water contaminated by *V. cholerae* can also be a source of infection. It is rare for cholera to be spread by contact with a cholera patient.

CLINICAL [3]

80% develop mild or moderate symptoms while approximately 20% develop acute watery diarrhea with severe dehydration, which can result in death if untreated.

Cholera can present in three stages [3].

(1) Mild diarrhea and vomiting, which progresses rapidly. The stools become watery, containing no fecal matter—"rice water stool." Abdominal pain and high fever are usually minimal or absent and distinctly disproportionate to the diarrhea. Electrolyte abnormalities may manifest as muscle weakness, severe abdominal cramps, intestinal ileus, or cardiac dysrhythmia [3]. The temperature rises but the skin is generally cold, blue, and the pulse is weak. The intake of water to quench thirst dilutes the body salts still further, and makes the cramps worse.

(2) Collapse, the body becomes colder, the skin dry, wrinkled, and purple. Dark, small outputs of urine or anuria may be present. Mental status changes secondary to hypoglycemia may also occur. Without urgent fluid and electrolyte replacement therapy, hypovolemic shock and death may occur as early as 24 h after the onset of the symptoms.

(3) The third stage. Recovery follows in favorable cases. The fluid loss decreases and there is improvement in the general condition. Note: a relapse may occur or the patient may sink into a condition resembling typhoid fever. Also, the condition may deteriorate over a period of 2 or 3 weeks. During this stage of reaction, the temperature may rise, and the patient may be in danger of developing pneumonia. The disease runs its course in 2–7 days. An important causes of death, in inadequately treated patients are hypovolemic shock, metabolic acidosis, and uremia resulting from acute tubular necrosis.

In epidemic situations, a clinical diagnosis is made by taking a history of symptoms from the patient and by clinical examination. Treatment is usually started without or before confirmation by laboratory analysis of specimens.

LABORATORY FINDINGS [3]

Stool and swab samples should be collected in the acute stage of the disease, before antibiotics have been administered. Direct examination of stool - can be useful for the presumptive identification of V. cholera.

Microscopy: Stool show straight or curved bacilli with rapid, darting, or "shooting star" motility suggestive of V. cholera infection.

Inhibition of the motility of the organism by *V. cholerae* O1 antiserum further supports the diagnosis of cholera.

Stool culture: Stool from suspected cholera patients should be cultured on a selective and differential media such as bile salt, GTT (gelatin-tellurite-taurocholate),TCBS (thiosulfate-citrate-bilesalt-sucrose) agar.

On bile salt or GTT agar, the organisms appear as typical translucent colonies within 24h.

On TCBS agar, V cholerae appear at 24h as distinct, large, flat yellow colonies.

Latex agglutination tests and ELISA-based methods to detect the presence of cholera toxin have been developed. Other tests include Polymerase chain reaction (PCR). These methods are rapid in assessing detection of *V. cholerae*.

RAPID DIAGNOSTIC TESTS FOR CHOLERA

SMART, Medicos Dip Stick, and an Institute Pasteur (IP) cholera dipstick tests have been developed.

The Institute Pasteur (IP) dipstick is the most appropriate rapid diagnostic assay for the detection of *V. cholerae* O1 in locations where the skill level of personnel may be low, such as in remote areas.

High cost may limit the utility of any diagnostic test in the developing world.

DIFFERENTIAL DIAGNOSIS [3]

Salmonella species, Shigella species, Yersinia enterocolitica, Campylobacter species and *E. coli*.

Viruses: rotavirus and Norwalk agent.

Parasites: giardia lambdia, Cyclospora, and Cryptosporidium.

ESSENTIALS OF DIAGNOSIS [3]

History of exposure, particularly travel to endemic or epidemic locales.

Acute onset of voluminous, watery diarrhea with low-grade fever and mild abdominal pain.

The presence of straight to curved gram-negative bacilli, with single polar flagellum, in the stool of an infected patient.

In wet preparation, these organisms demonstrate a characteristic darting or shooting star motility confirmed by motility inhibition with specific antisera.

Culture of *V. cholerae* from stool with differential media.

Bacterial growth in nutrient broth, without 1% Nacl supplementation.

This characteristic is useful for separating *V. cholerae* from most other Vibrio species.

Detection of V cholerae toxins by latex agglutination or enzyme-linked immunosorbent assay (ELISA) or the detection of V. cholera-specific nucleic acid by PCR-based methods.

TREATMENT

Oral or intravenous hydration is the mainstay of cholera treatment [4].

In conjunction with hydration, antibiotics are recommended for severely ill patients. Antibiotics should be administered particularly in patients who are moderately or severely dehydrated and continue to pass a large volume of stool during rehydration therapy. It is also recommended for hospitalized patients.

In most countries, doxycycline is recommended as first-line treatment for adults, while azithromycin is recommended for first-line treatment for children and pregnant women. During an epidemic or outbreak, antibiotic susceptibility regular testing of sample isolates should be done.

Education of health care workers, assurance of adequate supplies, monitoring of practices are important for appropriate dispensation of antibiotics [4].

Oral rehydration therapy (ORT) is a simple, cheap, and effective treatment for dehydration [3].

ORT consists of a solution of salts and sugars, which is taken by mouth. It is most important in the developing world, where it saves millions of children a year from death due to diarrhea—the second leading cause of death in children under five [3].

WHO ORS SOLUTION (2 L +/− 32 ML/KG) [3]

Some children develop worsening diarrhea and increasing serum sodium concentrations when treated with standard ORS solution and given only small amounts of plain water.

This is probably caused by the slight hypertonicity of standard ORS solution combined with transient partial glucose malabsorption. This can be avoided if water, breast milk, or another low-solute drink is given liberally during maintenance therapy with ORS solution, as recommended by the WHO.

The WHO and UNICEF jointly maintain the official guidelines for the contents of reduced osmolarity ORS packets. These guidelines are used by manufacturers of commercial ORS packets that are available for purchase. The reduced osmolarity ORS solution has a total osmolarity of 245 mmol/L.

ZINC SUPPLEMENTATION [3]

There is an additional recommendation of zinc supplementation for the management of diarrheal disease In addition to ORS, particularly for pediatric patients. For children under five, zinc supplementation significantly reduces the severity and duration of diarrhea and is strongly recommended as a supplement with ORS for dehydrated children. Preparations are available as a zinc sulfate solution for adults, a modified solution for children, and also a tablet form for children.

PROGNOSIS

Among people developing symptoms, 80% of episodes are of mild or moderate severity.

Among the remaining cases, 10%–20% develop severe watery diarrhea with signs of dehydration.

If untreated, as many as one in two people may die. Under ideal conditions and with prompt and adequate fluid replacement, mortality approaches zero, and significant sequelae are rare. Unfortunately, the prognosis for patients infected during cholera epidemic is often poor, due to lack of basic medical care, such as fluid and electrolytes replacement, which exists in situations that foster epidemic cholera. The success of treatment is significantly affected by the speed and method of treatment.

If cholera patients are treated quickly and adequately, the mortality rate is less than 1%; however, with untreated cholera the mortality rate rises to 50%–60%.

Patients who are not recognized to have an intestinal ileus may fare poorly. These patients may initially be misdiagnosed or if recognized to have cholera may not receive adequate fluid and electrolyte replacement.

PREVENTION AND CONTROL OF CHOLERA OUTBREAKS

Measures for the prevention of cholera consist of providing clean water and proper sanitation to populations potentially affected.

Health education and good food hygiene are important.

In endemic areas, water must be boiled before drinking and cooking. Vegetables and fruits should be washed with a solution of potassium permanganate before consumption. Other precautions include thorough washing of hands by all those who handle food, and elimination of all contacts with the disease.

The majority of patients—up to 80%—can be treated adequately through the administration of oral rehydration salts (WHO/UNICEF ORS standard sachet).

Very severely dehydrated patients are treated through the administration of intravenous fluids, preferably Ringer lactate. Appropriate antibiotics can be given to severe cases to diminish the duration of diarrhea, reduce the volume of rehydration fluids needed, and shorten the duration of vibrio excretion.

In order to ensure timely access to treatment, cholera treatment centers should be set up among the affected populations whenever feasible.

ORAL CHOLERA VACCINES [1]

Currently there are three WHO oral cholera vaccines (OCV)—Dukoral, Shankol, and Euvichol. All three vaccines require two doses for full protection. Dukoral can be given to all individuals over the age of 2 years. There must be a minimum of 7 days, and no more than 6 weeks, delay between each dose. Children aged 2–5 require a third dose. Dukoral is mainly used for travelers. Two doses of Dukoral provide protection against cholera for 2 years.

Shancol and Euvicol are essentially the same vaccine produced by two different manufacturers. They are given to all individuals over the age of 1 year. There must be a minimum of 2 weeks' delay between dose of these two vaccines. Two doses of either of these vaccines provide protection against cholera for 3 years, while one dose provides a short-term protection. The August 2017 WHO Position Paper on Vaccines Against Cholera states that OCV should be used in areas with endemic cholera, in humanitarian crises with high risk of cholera, and during cholera outbreaks; always in conjunction with other cholera prevention and control strategies. Also, vaccination should not disrupt the provision of other high-priority health interventions to control or prevent cholera outbreaks.

REFERENCES

[1] WHO. Cholera. 1 February, http://www.who.int/newsroom/fact-sheets/detail/cholera; 2018.
[2] Mukeredzi T. IRIN: Africa's all too preventable cholera crisis, https://www.irinnews.org/news/2018/02/01/Africa-s-all-too-preventtable-cholera-crisis; 2018.
[3] Paradwala H. Cholera. In: Segal I, Pitchumoni CS, Sung J, editors. Gastroenterology and hepatology manual: a clinicians guide to a global phenomenon. Australia: McGraw-Hill; 2011. [chapter 12].
[4] CDC. Antibiotic treatment. |Treatment| Treatment| Cholera |CDC. January 20, http://www.cdc.gov/cholera/treatment/antibiotic-treatment.html; 2015.

CHAPTER

CONCLUSIONS: PROBLEMS AND POSSIBLE SOLUTIONS

21

Isidor Segal

G.I.T. Unit, Chris Hani Baragwanath Hospital and University of Witwatersrand, Soweto, South Africa

*...Turning the epidemic*around required medical and nursing staff, African and international, to combine absolute discipline, courage, basic clinical skills and innovation that would have been admired by any generation of physicians*

C.M. Whitty: Harveian Oration-2017 [1]

It is compelling that recent research is revealing an unpredictable diversity of emerging digestive disease patterns in sub-Saharan Africa.

In LMICs, social behavior is recognized as a leading cause of ill health through excessive use of alcohol and smoking. These common promoters of cardiovascular, respiratory diseases and cancer often result in premature death. Significant etiological factors also indicate that alcohol is a decisive negative factor in digestive disease, particularly esophageal, pancreatic, and liver diseases.

The major cancer in African men is carcinoma of the esophagus. Evidence suggests that changes may occur through nutritional deficiencies, promoted by the effects of alcohol, rather than the direct impact of a chemical carcinogen. Another risk factor in South African Blacks is pipe tobacco, smoked either in hand-rolled cigarettes or in pipes.

Heavy alcohol consumption is again one of the three common environmental factors in acute and chronic pancreatitis and two further environmental components result from exposure to occupational chemicals and a very low intake of fruit. It is apparent that "healthy" Sowetans are in a state of oxidant stress that is linked to poor Vitamin C status and low selenium levels. These findings offer research scope for disease prophylaxis in underprivileged communities.

The fact that alcohol and smoking promote cancer is a signal that community involvement is necessarily a major player, as underlined by the success of Adherence Community Clubs, in helping to combat HIV/AIDS.

The urbanization process is strongly associated with poverty. More than two-thirds of urban dwellers live in slums, their lives driven by extremely poor living conditions and a lack of basic services: no clean water, inadequate sanitation, unacceptable unhygienic conditions as experienced in an overcrowded, bereft environment.

Overriding the various diseases is always the theme of poverty.

Solutions relate to improved governance and job creation is essential to eradicate this phenomenon. Funds for health initiatives and health care access are priorities for changing the spectrum of preventable diseases in SSA.

*West African Ebola epidemic

Digestive Diseases in Sub-Saharan Africa. https://doi.org/10.1016/B978-0-12-815677-3.00021-9

The impact of rapid urbanization continues to encroach on nutritional transition resulting in epidemiological changes. This is characterized by an unpredictable increase in overnutrition, resulting in overweight, obesity, and other noncommunicable diseases and concomitantly, there is a decrease in undernutrition and infectious disease.

The dawn of the 21st century is witnessing green shoots that hint at subtle environmental changes; recognition of natural resources, upward mobility, and demographic movements that are beginning to influence changes in the digestive disease profile.

The limited data available on GERD, Hiatus Hernia, and Barrett's esophagus suggest that these diseases are rare in SSA. However, there are no data on these diseases and undoubtedly there may be a potential bias in reporting. When compared with other regions of the world, treatment regarding access to care, satisfying patient expectations, and willingness to undergo procedures is absent.

Most patients present at clinics, rather than hospitals for what they often consider as trivial complaints. Thus, careful records become crucial to clinics for patient follow-up of GERD. It is essential to share the information with hospitals for more consistent and positive results. The scope of clinics, as the first port of call for the sick, should emphasize the importance of continuity with patients. Records, diagnosis, and treatment of patient records should be available to the clinics and hospitals. Clinical services need to be upgraded to accommodate to contemporary technology since health personnel and many patients have IT devices they can readily be contacted for information and confirmation.

Prevalence of *H. pylori* infection in adults and children is high. It is a pathogen in peptic ulcer disease and a Class 1 carcinogen, present in most GU and DU patients. Variations in some African countries report high H. pylori prevalence with a low rate of gastric cancer.

The African Enigma has not yet been resolved and the solution requires further research.

Many studies show a relationship between NSAIDS and GU, indicating adverse consequences that are not made explicit. Greater spending power in an urban milieu will inevitably lead to an increased use of NSAIDS with concomitant increases in GU.

Interestingly, DU experiments show that protective activity lies in the phospholipid and sterol fractions of unmilled staples.

Urbanization has led to a change from home pounding to milled white maize, white bread, and white rice as a staple diet. Nutritional advice on the benefits of unmilled maize and rice may have positive effects in increasingly aware urban populations

During WW2, large bowel disease patterns changed as a result of national diets with rigid food restrictions. Segal and Walker noted an unusual outcome, during this period the frequency of noninfective bowel diseases had decreased.

Large bowel diseases: IBD is uncommon in African Black populations but recently the incidence is increasing. The prevalence in Caucasian populations in South Africa is similar to that of HICs. It is in Colored [Mixed Race] populations however that a significant rise in IBD, particularly Crohn's disease, [CD] has occurred. CD manifests complications that highlight the relationship of TB and CD. TB is a great mimic of other diseases and a high index of suspicion is always necessary for a correct diagnosis.

Colorectal Cancer (CRC) is increasingly being diagnosed in young Africans compared with older Caucasians. Once again, deprivation is responsible for the rise of CRC, and this also explains why genetics and genetic syndromes have not been addressed in SSA.

Three other diseases that thrive in poverty are intestinal schistosomiasis, amoebiasis, and cholera. An awareness of clean water is a crucial factor in the prevention of cholera, amoebiasis, and other infections. Political indifference has contributed to the high prevalence of disease with a stable government and positive health strategies, eradication of disease could begin in Africa.

The commonest colonic disease in HICs is diverticular disease (DD). It was uncommon in LMICs but following changes in dietary patterns DD have emerged and are becoming more prevalent.

Liver disease's burden is compounded by lack of access to essential medical, social, and financial resources. Hepatitis A (HAV) is endemic as a consequence of extreme poverty. The tragedy of Hepatitis B (HBV) that is currently responsible for 70% of hepatocellular cancers is a cancer that is preventable with a birth dose of vaccine. All carriers of HBV however would require close monitoring and follow up on therapy.

To eradicate Hepatitis C (HCV) global leadership and education, in basic hygiene, is vital to preventative treatments. Proper funding for antiviral drugs could offer an almost universal cure within a short duration.

Hepatocellular cancer (HCC) is the second most common cancer in men, with viral hepatitis as the commonest risk factor, reflected in the high rate of chronic HBV. The incidence has been underestimated, compounded by lack of availability and inaccurate registry data. Nevertheless, it is believed that the next few decades will herald dramatic reductions in the prevalence of HCC.

Significant challenges remain that rely on government commitment to effective health care interventions to stem the flood of cancer in Africa.

Biliary disease has become common in South Africa. The disease is indicative of the changing demography of diseases in SSA, with late presentation. A problem that increases morbidity and mortality and with rapid urbanization there are limited databases on the continent to accurately reflect the spectrum of the disease. Expertise is however being developed and attention paid to address the many issues.

HIV is an example of a new disease that has ravished Africa. Once the disease was better understood, a combination of life-changing factors and successful antiretroviral therapy began reversing the trend. With the assistance of pharmaceutical companies, media, and community working together, the "end game" is in sight.

Some thoughts on ideas for dealing with difficult often unusual circumstances to improve the health crisis SSA are considered here.

Health campaigns are being driven to promote health awareness. To this end, nutritional education focused on healthy food as part of the curriculum for schools, universities are underway. Media has accepted a responsible role assisted by counselors and health workers.

Adherence clubs, community support groups with Traditional Healers are integral in providing a bridge for the conventional western health system. They will facilitate the role of psychologists and encourage clients to attend clinics and hospitals when necessary contributing to preventive health issues.

To accomplish the realization of these expectations, leadership skills are necessary since international assistance grants are urgently needed for training and establishing bilateral exchange programs. Donations of all basic medical equipment, new or old, physical mechanical aids, IT components of all sorts, books, journals, and ranges of ingenuity using unknown elements will contribute to the way forward.

A good example initiated by the Royal College of Physicians is the "Physicians for Africa" program, i.e., partnering with leading physicians in East, Central, and Southern Africa to provide a new professional training college, designed to ameliorate the situation in these countries where access to doctors is severely limited.

Focus for advances in treatment for digestive disease in LICs would include Hepatitis C, where several new classes of therapies are now available that are a marked improvement on those of past treatments. Technology and diagnostic methodology including stem cell research are changing the face of conventional treatment, aided by the progression of communication facilities. Improved workforce planning is underway to stem the migration flow of professionals from their own countries to richer countries globally.

Health infrastructure in Africa to be relevant in the 21st century requires a strong, viable foundation to take care of its own people and contribute to world medicine.

In conclusion, an extraordinary precedent witnessed by the recent discovery of rescuing the vast medieval library of manuscripts (including pharmacology and astronomy) from Timbuktu, the famed ancient capital of central Africa. This was achieved only through the committed co-operation of African and International sources, a signpost to the future of sub-Saharan African [2].

The ultimate aim is the establishment of centers of excellence and the creation of health infrastructure to lead and ensure world-class standards of medical expertise and care in sub-Saharan Africa (Tables 21.1–21.6).

Table 21.1 Comparison of Traditional Versus Emergent GIT Diseases

	Incidence of Traditional GIT Diseases				Emergence of GIT Diseases			
	0	+	++	+++	0	+	++	+++
Esophageal cancer				+++				+++
GERD	0				0			
Gastric ulcer		+					++	
Gastric cancer	0					+		
Duodenal ulcer	+/0	+/0				+/++	+/++	
Pancreatitis	0							+++
Appendicitis		+				+		
Sigmoid volvulus				+++		+		
IBD	0					+		
Diverticular disease	0						++	
Colorectal cancer	0					+		
Viral hepatitis				+++				+++
Liver cancer				+++			++	
Biliary disease	0							+++

Table 21.2 Ramifications of Nutritional Changes on GIT Disease

Cause	Effect/s
Micronutrient deficiencies	Esophageal cancer; Pancreatitis
Change from unrefined diet to refined diet	Diverticular disease; Diabetes; Biliary disease; Duodenal ulcer; Colorectal cancer; Obesity
High salt diet	Hypertension
Under nutrition	
High unrefined diet	Sigmoid volvulus

Table 21.3 Complications of Alcohol Consumption

Complications of alcohol consumption
Behavior disturbances
Esophageal cancer
Pancreatitis
Cirrhosis of liver

Table 21.4 Emerging Nonintestinal Diseases in SSA

Emerging nonintestinal diseases in SSA
Obesity
Diabetes mellitus
Mental illness
Cardio-vascular disease
Extra-intestinal cancers

Table 21.5 Urbanization of Poverty

Urbanization of poverty
Slums
Poor basic services
Lack of water supply
Sanitation
Open defecation
Forbes and Lindfield

Table 21.6 Urbanization

Urbanization
↓
Upward mobility
Demography – In 2050 +/- 20% of the world's population in Africa
Xenobiotics - Smoking
Alcohol
Air pollution
Climate change
Dietary changes

REFERENCES

[1] Whitty JM. Harveian Oration 2017: triumphs and challenges in a world shaped by medicine. Clin Med 2017;17(6):537–44.

[2] English C. The book smugglers of Timbuktu. London: William Collins; 2017.

Index

Note: Page numbers followed by *f* indicate figures, *t* indicate tables and *b* indicate boxes.

T

9780128156773

9780128156773